Crafting Superior Environmental Enforcement Solutions

William L. Thomas,

Bertram C. Frey,

and Fern Fleischer Daves

Environmental Law Reporter
Environmental Law Institute
Washington, D.C.

1616 P Street NW, Washington, DC 20036

Published August 2000.

Printed in the United States of America
ISBN 1-58576-009-9

Cover art: Francisco de Goya y Lucientes (1746-1828), *The Forge*,
The Frick Collection, New York.

Library of Congress Cataloging-in-Publication Data

Thomas, William L.
Crafting superior environmental enforcement solutions /
William L. Thomas, Bertram C. Frey, and Fern Fleischer Daves.
p. cm.
Includes bibliographical references.
ISBN 1-58576-009-9
1. Environmental law—United States—Criminal provisions.
2. Compromise (Law)—United States. 3. Environmental engineering.
I. Frey, Bertram C. II. Daves, Fern Fleischer. III. Title.

KF3775.T49 2000
344.73'046—dc21 00-061685

Crafting Superior Environmental Enforcement Solutions

William L. Thomas,

Bertram C. Frey,

and Fern Fleischer Daves

Table of Contents

Foreword iv

Article:

Using Auditing, Pollution Prevention, and Management Systems to Craft Superior Environmental Enforcement Solutions, by William L. Thomas, Bertram C. Frey, and Fern Fleischer Daves 3

Appendices:

Appendix A: State Audit Privilege and Immunity Legislation and Self-Disclosure Policies 23

Appendix B: Audit Provisions in Federal Court Consent Decrees or Orders 25

Appendix C: Audit Provisions in EPA Administrative Orders 59

Appendix D: Pollution Prevention Supplemental Environmental Projects in Fiscal Year 1999 65

Appendix E: Decrees, Orders, and Settlements Requiring Features of a Compliance-Focused Environmental Management System 73

Decrees and Agreements:

Consent Decree (U.S. v. ASARCO Inc.) 101

Consent Decree (U.S. and Texas v. Encycle/Texas, Inc. and ASARCO Inc.) 113

Plea Agreement (U.S. v. BP Exploration (Alaska) Inc.) 131

Bibliography 145

Foreword

Crafting Superior Environmental Enforcement Solutions examines the creative ways in which prosecutors, regulators, and industry have used environmental auditing and pollution prevention techniques to resolve enforcement actions. In particular, Supplemental Environmental Projects (SEPs) and Environmental Management Systems (EMSs) are discussed, with examples provided of their use.

The book additionally provides detailed analyses of (1) state audit privilege and immunity statutes and policies, (2) environmental audit provisions in federal consent decrees and in EPA administrative orders, (3) pollution prevention SEPs approved in 1999, and (4) other instances of EMS commitments at the federal and state level. It also includes copies of important agreements reached between the government and private parties regarding the use of EMSs. The book supplements the discussion contained in Frank Friedman's well-known and utilized text, *Practical Guide to Environmental Management* (8th ed.), an ELI publication that in its most recent edition has been extensively revised and expanded.

Crafting Superior Environmental Enforcement Solutions is yet another example of the ELI Publications group's close attention and hard work. The book was edited by John H. Turner, Vice President, Publications and *ELR* Editor-in-Chief. In addition, *ELR* Managing Editor Linda L. Johnson did an outstanding job in preparing the book, and special thanks go to Editorial Assistant Carolyn Fischer and Desktop Publisher William J. Straub. Kim Goldberg is responsible for marketing of the book.

As a national environmental research and publishing organization dedicated to the development of more effective and more efficient environmental protection efforts, ELI is pleased to present *Crafting Superior Environmental Enforcement Solutions*. We hope environmental lawyers, managers, corporate decisionmakers, and agency personnel will find it invaluable.

—J. William Futrell, President
Environmental Law Institute

Using Auditing, Pollution Prevention, and Management Systems to Craft Superior Environmental Enforcement Solutions

by William L. Thomas,

Bertram C. Frey,

and Fern Fleischer Daves

Using Auditing, Pollution Prevention, and Management Systems to Craft Superior Environmental Enforcement Solutions

by William L. Thomas, Bertram C. Frey, and Fern Fleischer Daves

"'Now, that's what I was afraid of!' Porfiry exclaimed hotly and as if involuntarily.

'That's what I was afraid of, that you don't want our reduction.'"

—Fyodor Dostoevsky, *Crime and Punishment*[1]

The "environmental cop"[2] is very much on the beat, requiring environmental cleanup and meting out punishment for transgressions of U.S. law. In fiscal year (FY) 1999, the U.S. Environmental Protection Agency (EPA) concluded enforcement cases requiring a record $3.6 billion for environmental cleanup, pollution control equipment, pollution prevention and improved monitoring; a record $166.7 million in civil and administrative penalties; and $61.6 million in criminal fines. The Agency took 3,935 civil and administrative enforcement actions and referred 241 cases for criminal prosecution.[3] And, as in past years,[4] guards escorted corporate employees and managers to prisons which, unlike those etched by Piranesi, are anything but imaginary.[5]

Giovanni Battista Piranesi, *The Drawbridge*, 1761

Some within the enforcement community have, however, begun to shift the focus from pure punishment and deterrence to the use of strategic environmental management to address the root causes of historic noncompliance in selected cases.[6] For many of the companies involved in these agreements, the result has been a marked reduction of enforcement penalties and stronger environmental perfor-

Mr. Thomas is an associate in the Washington, D.C., office of Winthrop, Stimson, Putnam & Roberts. Mr. Frey is deputy regional counsel in U.S. Environmental Protection Agency's (EPA's) Region 5 office. Ms. Fleischer Daves, an attorney in private practice, was corporate environmental counsel of ASARCO Incorporated from September 1992 to January 2000. Although Ms. Fleischer Daves provided information on the agreements negotiated between ASARCO and the government during her tenure as corporate environmental counsel for ASARCO, Mr. Thomas wrote the portions relating to the ASARCO consent decrees. Any opinions expressed are those of the authors and do not represent an endorsement of any point of view by EPA.

1. Fyodor Dostoevsky, Crime and Punishment 459 (Richard Pevear & Larissa Volokhonsky trans., Alfred A. Knopf, Inc. 1992) (1866). To a limited extent, this Monograph revisits themes touched upon by William N. Farran III & Thomas L. Adams Jr., *Environmental Regulatory Objective: Auditing and Compliance or Crime and Punishment*, 21 ELR 10239 (May 1991).

2. We borrow this characterization from Sylvia Lowrance, as quoted in Cheryl Hogue, *Record $76.7 Million in Criminal Fines Leads EPA Accomplishments for Fiscal 1996*, Daily Rep. for Executives (BNA), Feb. 26, 1997, at A-27, A-29.

3. U.S. EPA, Office of Enforcement and Compliance Assurance, Highlights of FY 1999 EPA Enforcement and Compliance Assurance Activities (Jan. 19, 2000) [hereinafter 1999 Highlights]. Compare these results with FY 1998, when the Agency made over 677 criminal and civil cases with over $180 million assessed in penalties and fines; the criminal program set annual records for defendants charged (350) and investigations initiated (636). U.S. EPA, Office of Enforcement and Compliance Assurance, FY98 Accomplishments Report (June 1999).

4. Defendants were sentenced to a record 208 years of prison time for environmental crimes in 1999. *See* 1999 Highlights, *supra* note 3. For press accounts of recent cases involving prison time for violators, *see, e.g.*, Steve Cook, *Businessman Hit With 17-Year Sentence, $6 Million Penalty for Cyanide Tank Incident*, Daily Env't Rep. (BNA), May 2, 2000, at AA-1; *President of Consulting Firm Sentenced to Prison for Waste Water Injection Scheme*, 31 Env't Rep. (BNA) 698 (Apr. 14, 2000); John H. Stam, *Court Upholds Criminal RCRA Conviction; Chemical Salesman Must Serve 56 Months*, 30 Env't Rep. (BNA) 1279 (Nov. 12, 1999); *Florida Businessman Receives 13 Years in Record Sentence Under Federal Statutes*, 30 Env't Rep. (BNA) 800 (Aug. 20, 1999); Barney Tumey, *Georgia Chemical Plant Executives Receive Prison Terms for Violations*, 30 Env't Rep. (BNA) 319 (June 18, 1999); Tripp Baltz, *Court Imposes 14-Year Prison Sentence in Colorado Hazardous Waste Disposal Case*, 30 Env't Rep. (BNA) 208 (June 4, 1999); *Ninth Circuit Approves Four-Year Sentence for Unlawful Disposal of San Diego Sludge*, 29 Env't Rep. (BNA) 2477 (Apr. 16, 1999); *Ninth Circuit Approves Prison Term for Executive Who Violated Water Act*, 29 Env't Rep. (BNA) 1699 (Jan. 1, 1999); *Prison Sentence Given for Illegal Transport of Waste*, 28 Env't Rep. (BNA) 2748 (Apr. 24, 1998); *Businessmen Sentenced for Fraud, RCRA Violations*, 28 Env't Rep. (BNA) 1671 (Jan. 2, 1998). Much has been written about the criminalization of environmental law. One might well begin with the series of articles by the criminal defense team at Venable, Baetjer, Howard & Civiletti, *see* John F. Cooney et al., *Criminal Enforcement of Environmental Laws: Parts I-III*, 25 ELR 10459, 10525, and 10600 (Sept., Oct., and Nov. 1995, respectively). For a more recent perspective, see Joe D. Whitley & Douglas S. Arnold, *Doing Time for Gree Crimes: Watch Out for Heavy Fines and Long Prison Terms for Corporate Officers*, Legal Times, May 22, 2000, at 32; Judson W. Starr & Kathleen S. Dolan, *The Increase of Criminal Exposure in the American Workplace*, Nat. Resources & Env't, Spring 2000, at 239; Judson W. Starr et al., Criminal Enforcement of Environmental Laws (American Law Institute 2000).

5. Of the various images in Giovanni Piranesi's 1761 collection of etchings from *Carceri D'Invenzione* (Imaginary Prisons), one of the most extraordinary is "The Drawbridge," which appears on page 60 of Giovanni Battista Piranesi, The Complete Etchings (John Wilton-Ely ed., Alan Wolfsky Fine Arts 1994) (Vol. I) (Fine Arts Museums of San Francisco, Achenbach Foundation for Graphic Arts purchase, 1969.32.7.7).

6. *See* Steven Sisk, Overview of EMS and Enforcement Training Class, Presentation at the Multi-State Working Group on Environmental Management Systems Quarterly Meeting, Oakland, Cal. (Dec. 1999).

mance over time. Perhaps Plato was right when he observed in *Georgias* that "punishment brings wisdom."[7] This Monograph will examine the creative manner in which prosecutors, regulators, and industry have used environmental auditing, pollution prevention techniques, and environmental management systems to resolve enforcement actions.

Environmental Auditing and Prosecutorial Leniency

The discipline of environmental auditing, pioneered by American chemical and petrochemical firms in the early 1970s to ensure compliance with pollution legislation, is now a common element of the environmental management systems of most large firms in the United States, and increasingly, Europe.[8] Generally defined as a systematic, documented, periodic, and objective review by regulated entities of facility operations and practices related to meeting environmental requirements,[9] an audit may have one or multiple purposes. An auditor might, for example, evaluate a facility's compliance with environmental requirements, the efficacy of the firm's environmental management system (EMS), or the nature and extent of risks to the firm (and relevant stakeholders) from regulated and unregulated materials and practices.[10] Larger organizations generally have greater resources to devote to an internal auditing team, while smaller entities are more likely to use outside auditors. Whether performed by independent, internal, or third parties, however, voluntary private auditing efforts inform corporate strategy and decisionmaking, and complement regulatory agency inspections and other forms of governmental oversight. The fact that audit materials also serve as a convenient road map for prosecutors has also given rise to vigorous debate over the prudence of such self-examinations and whether some sort of formal legal privilege ought to cover them. EPA, resisting efforts of many states that have created statutory shields protecting audit documentation from disclosure,[11] has sought to promote the use of self-evaluation through a series of carefully crafted nonbinding guidance documents. These policies have had a marked impact on the negotiation and outcome of enforcement disputes.

EPA's 1986 Audit Policy

It is EPA policy to encourage the use of environmental auditing by regulated entities to help achieve and maintain compliance with environmental laws and regulations, as well as to help identify and correct unregulated environmental hazards. Recognizing that voluntary auditing programs can result in better identification, resolution, or avoidance of environmental problems, the Agency issued its first policy on the subject in July 1986. Generally, the policy: (i) encourages regulated entities to develop, implement, and upgrade environmental auditing programs; (ii) discusses when the Agency may or may not request audit reports; (iii) explains how EPA's inspections and enforcement activities may respond to regulated entities' efforts to assure compliance through auditing; (iv) endorses environmental auditing at federal facilities; (v) encourages state and local environmental auditing initiatives; and (vi) outlines elements of effective auditing programs.[12]

EPA has broad statutory authority to request relevant information on the environmental compliance status of regulated entities.[13] The Agency has elected not to avail itself of this authority, believing routine requests for audit reports could inhibit auditing in the long run, decreasing both the quantity and the quality of audits conducted.[14] But EPA and state and local environmental regulators may seek audit reports during investigations that precede an enforcement action. Private groups, including citizens environmental groups, may also seek disclosure during discovery in citizen suits and toxic tort litigation. In addition, such documents are often sought by insurance companies in coverage litigation related to environmental damage claims.

Under the 1986 Audit Policy, EPA did not promise to forgo inspections, reduce enforcement responses, or offer other incentives in exchange for implementation of environmental auditing or other sound environmental management practices. Although environmental audits complement inspections, the consensus within the Agency seems to be that such self-assessments are no substitute for regulatory oversight. In addition, certain EPA programs, including the Resource Conservation and Recovery Act (RCRA) hazardous waste regime, require minimum facility inspection frequencies to which EPA and the states and owners and operators of regulated facilities must adhere.[15]

While EPA inspections of self-audited facilities will therefore continue, to the extent that compliance performance is considered in setting inspection priorities, facilities with a good compliance history may be subject to fewer inspections.[16] Moreover, in fashioning enforcement re-

7. Plato, Collected Dialogues 229, 262 (Edith Hamilton & Huntington Cairns eds., Princeton Univ. Press 1961) (W.D. Woodlead trans.).

8. *See generally* Kevin J. Sobnosky, *The Value-Added Benefits of Environmental Auditing*, Envtl. Quality Mgmt., Winter 1999, at 25; Air Emissions, Baselines, and Environmental Auditing (Jacqueline Shields ed., John Wiley & Sons 1997); Frances Cairncross, Green Inc. 201-03 (Island Press 1995); Richard Welford & Andrew Gouldson, Environmental Management and Business Strategy 100 (Pitman 1993); Frances Cairncross, Costing the Earth 292-97 (Harvard Bus. School Press 1991). Commentary and guidance on environmental auditing abounds. For a sample, *see* Frank B. Friedman, Practical Guide to Environmental Management (8th ed., Environmental Law Inst. 2000); Bruce W. Piasecki et al., Environmental Management and Business Strategy: Leadership Skills for the 21st Century (John Wiley & Sons 1999).

9. U.S. EPA, Environmental Auditing Policy Statement, 51 Fed. Reg. 25003, 25006 (July 9, 1986), Admin. Mat. 35001 (available from the ELR Document Service, ELR Order No. AD-3216) [hereinafter 1986 Audit Policy]. The International Organization for Standardization (ISO) has essentially adopted EPA's definition of environmental auditing. *Compare* 1986 Audit Policy, at 25004, 25006-07 *with* ISO 14010. On the ISO 14000 auditing standards, *see generally* John C. Shideler, *What Are the Remaining Issues of ISO 19011?*, Int'l Envtl. Sys. Update, Jan. 2000, at 8; Jason Hart, *The Evolution and Revolution of Integration: New Common Auditing Document Will Make Cents for Companies*, Bus. Standards, Nov./Dec. 1999, at 13.

10. 1986 Audit Policy, *supra* note 9, at 25006.

11. *See* discussion of state audit laws *infra* notes 47-49.

12. 1986 Audit Policy, *supra* note 9, at 25004.

13. *E.g.*, 33 U.S.C. §1318, ELR Stat. FWPCA §308; 42 U.S.C. §7414, ELR Stat. CAA §114; 42 U.S.C. §6927, ELR Stat. RCRA §3007.

14. 1986 Audit Policy, *supra* note 9, at 25007.

15. *Compare* 42 U.S.C. §6927(e), ELR Stat. RCRA §3007(e) *with* 42 U.S.C. §6924(a)(2), ELR Stat. RCRA §3004(a)(2) *and* 40 C.F.R. §264.15 (1999).

16. Under U.S. EPA's National Environmental Performance Track Program, launched June 26, 2000 (see <http://www.epa.gov/performancetrack/about/descriptionjune2000.pdf>), "top environmental performers," i.e., "those who go beyond compliance with regulatory requirements to attain levels of environmental performance and management that

sponses to violations, it is EPA policy to take into account, on a case-by-case basis, the honest and genuine efforts of regulated entities to avoid and promptly correct violations and underlying environmental problems.[17] When regulated entities take reasonable precautions to avoid noncompliance, expeditiously correct underlying environmental problems discovered through audits or other means, and implement measures to prevent their recurrence, EPA may (and often does) exercise its discretion to consider those actions as honest and genuine efforts to assure compliance.[18] This is particularly the case when a regulated entity promptly reports violations or compliance data which otherwise were not required to be recorded or reported to EPA.[19]

Nonetheless, it remains the case that the federal government has protected prosecutorial discretion first, leaving the fate of audit reports to be decided by individual enforcement and prosecution teams based upon their assessment of the circumstances. The 1986 Audit Policy expressly reserves EPA's right to request audit summaries or reports on a case-by-case basis whenever: (i) audits are conducted under consent decrees or other settlement agreements requiring them; (ii) a company has asserted its environmental management practices in mitigation or as a defense; or (iii) state-of-mind or intent is a relevant element of inquiry, such as during a criminal investigation.[20] The 1986 Audit Policy remains in effect. Some provisions have, however, been superceded by subsequent policy guidance as described further below.

U.S. Department of Justice "Factors"

On July 1, 1991, the U.S. Department of Justice (DOJ) issued a policy according limited consideration to environmental audits. This policy, entitled *Factors in Decisions on Criminal Prosecutions for Environmental Violations in the Context of Significant Voluntary Compliance or Disclosure Efforts by the Violator*, or simply *Factors*,[21] states that in determining whether to prosecute, DOJ attorneys should consider if the target has made "voluntary, timely, and complete disclosure." The policy offers a range of helpful examples of such cooperation, but essentially places the target at the mercy of the prosecutor. This policy has been invoked in a number instances to prevent a criminal prosecution, but the quid pro quo has almost always been a comprehensive and expensive civil settlement.

EPA's 1995 Self-Policing Policy

After seeking input from industry and other stakeholders over an 18-month period, and to offer an alternative to state and federal legislative efforts to enact environmental audit privilege and/or immunity laws, EPA announced a new policy dealing in part with environmental auditing on December 18, 1995.[22] More of a self-policing policy than an audit policy, the document was designed to "enhance the protection of human health and the environment" and encourage greater compliance with environmental laws and regulations.[23] The nonbinding policy was an attempt to offer an incentive for those firms that voluntarily discover and disclose violations, and can satisfy certain conditions.

The 1995 Self-Policing Policy having been described in detail by other commentators,[24] a brief summary should suffice. First, gravity-based penalties, i.e., for any non-economic benefit, may be eliminated when violations are found through either an environmental audit or a compliance monitoring program and are voluntarily and promptly disclosed and corrected. The environmental audit must be systematic, objective, and periodic, as defined under the 1986 Audit Policy. The compliance monitoring system must also meet certain requirements. These include standards and procedures that outline how employees are to meet compliance, a method for overseeing compliance, mechanisms to assure compliance (including monitoring and audits to detect and correct violations), efforts to communicate standards to employees and agents, incentives to managers and employees to comply with policies, proce-

benefit people, communities, and the environment" will be given a low priority for inspection targeting purposes. *See* Pam Parry, *U.S. EPA Launches Nationwide Incentives Program*, Int'l Envtl. Sys. Update, July 2000, at 8.

17. 1986 Audit Policy, *supra* note 9, at 25007.

18. *Id.*

19. *Id.*

20. *Id.* at 25007-08.

21. Admin. Mat. 35399 (available from the ELR Document Service, ELR Order No. AD-505).

22. U.S. EPA, Incentives for Self-Policing: Discovery, Disclosure, Correction and Prevention of Violations, 60 Fed. Reg. 66705 (Dec. 22, 1995), Admin. Mat. 35639 (available from the ELR Document Service, ELR Order No. AD-3125) [hereinafter 1995 Self-Policing Policy]. For further information on the policy, *see* <http://www.es.epa.gov/oeca/ore/apolguid.html> and <http://www.epa.gov/region5/ore/audits/audit_resources.htm>. The 1995 Self-Policing Policy applies to the assessment of penalties for any violations under all of the federal environmental statutes that EPA administers, and supercedes any inconsistent provisions in media-specific penalty or enforcement policies and EPA's 1986 Audit Policy. To the extent that existing EPA enforcement policies are not inconsistent, they will continue to apply in conjunction with this policy. *Id.* at 66712.

23. Similar policies were issued for small businesses, *see, e.g.*, U.S. EPA, Policy on Compliance Incentives for Small Businesses, 61 Fed. Reg. 27984 (June 3, 1996), Admin. Mat. 35649 (available from the ELR Document Service, ELR Order No. AD-2941), *superceded by* U.S. EPA, Small Business Compliance Policy, 65 Fed. Reg. 19630 (Apr. 11, 2000) (the revised Small Business Policy is available at <http://es.epa.gov/oeca/smbusi.html>, and small communities, *see, e.g.*, Memorandum from Steven A. Herman, Assistant Administrator, EPA, to Assistant Administrators, General Counsels, Regional Administrators, Deputy Regional Administrators, Regional Counsel and Regional Enforcement Coordinators, regarding Policy on Flexible State Enforcement Response to Small Community Violations (Nov. 22, 1995).

24. For additional discussion of, and reaction to, the 1995 Self-Policing Policy and its interim antecedent, 60 Fed. Reg. 16875 (Apr. 3, 1995), *see generally* Debra Hunter Johnson & Brian King, Voluntary Self-Disclosure and the Audit Policy—An In-House Perspective, American Bar Association (ABA), Section of Natural Resources, Energy, and Environmental Law, Annual Conference on Environmental Law, Keystone, Colo. (Mar. 2000); John Voorhees & Robert A. Woellner, International Environmental Risk Management: ISO 14000 and the Systems Approach 115-20 (Lewis Publishers 1998); Lynn L. Bergeson & Lisa M. Campbell, *Carrot or Stick: The Debate Over Environmental Auditing*, Wash. Law., Sept./Oct. 1997, at 38 [hereinafter Bergeson & Campbell]; Ellen Page Delsole, *An Environmental Audit Privilege: What Protection Remains After EPA's Rejection of the Privilege?*, 46 Cath. U. L. Rev. 325 (1997); John S. Guttman & Holly Cannon, *Eyes on the Green*, Legal Times, May 13, 1996, at 41; James T. Banks, *EPA's New Enforcement Policy: At Last, a Reliable Roadmap to Civil Penalty Mitigation for Self-Disclosed Violations*, 26 ELR 10227 (May 1996); Jim Gribble, *Can Anything You Find Be Used Against You?*, Bus. J. (Milwaukee), Mar. 2, 1996, at 21; David J. Freeman & Gregory R. Belcamino, *EPA's Audit Policy: Half a Loaf Is Better Than None*, N.Y. L.J., Feb. 15, 1996, at 1; *State Officials Offer Support for EPA's Final Audit Policy*, Inside EPA Wkly. Rep., Jan. 12, 1996,

dures for prompt and appropriate correction of any violations, and measures to prevent future violations. Gravity-based penalties may be reduced by up to 75 percent when a violation is voluntarily discovered, promptly disclosed, and expeditiously corrected, even if it was not found through an environmental audit or a compliance management system. As additional incentives, EPA pledges that if the conditions of the 1995 Self-Policing Policy are met, it will not recommend criminal prosecution and will refrain from routine requests for audits. If the Agency has independent evidence of a violation, it may seek additional information.

In order to receive full benefits under the 1995 Self-Policing Policy, a violator must meet several conditions. The entity must, for example, discover the violation through an environmental audit or compliance monitoring system "due diligence." The violation must be discovered voluntarily, i.e., the violation cannot be one discovered through legally mandated monitoring or sampling requirements prescribed by statute, regulation, permit, judicial or administrative order, or consent agreement. The disclosure must be prompt[25] and must be made in writing to EPA. The violation must be discovered and disclosed prior to any government investigation, citizen suit, filing of a third-party complaint, report of violation by an employee not in authority, or imminent discovery by a regulatory agency. The entity must correct the violation within 60 days, and remedy any harm that has occurred as a result of the violation. If it will take longer than 60 days to correct, the entity must notify EPA in writing. The regulated entity also must prevent recurrences of violations, which may require improvements to environmental auditing or "due diligence." The specific violation must not have occurred within the past three years at the same facility, or be part of a pattern of violations by the company's parent organization in the past five years. The entity must cooperate fully with EPA to determine the applicability of the policy. No penalty reduction will be available for companies whose violations "resulted in serious actual harm or which may have presented an imminent and substantial endangerment to public health or the environment." Corporations remain criminally liable for violations that exhibit an ongoing management philosophy or practice that conceals and/or condones violations, or high-level management's participation in (or willful blindness to) the violation.

In order to assure national consistency and in response to industry concern over apparent inconsistent and untimely enforcement review of self-disclosures under this policy, EPA established an Audit Policy Quick Response Team (QRT). The QRT is made up of senior representatives from EPA Headquarters, Regions and the DOJ. In January 1997, the QRT developed an Audit Policy Interpretive Guidance covering a range of common questions and issues and setting forth guidance for EPA's use in exercising its enforcement discretion.[26] EPA also publishes a periodic newsletter to inform the public of news and developments under the policy.[27] EPA established a separate Voluntary Disclosure Board (VDB) in October 1997, to act as a central body for reviewing all possible criminal violations disclosed under the policy. Like the QRT, the VDB's primary purpose is to ensure consistent policy application nationwide.

In May 1998, EPA embarked on a study of the "effectiveness" of the 1995 Self-Policing Policy. The Agency began by publishing a survey seeking information on how the policy and its application could be improved.[28] The Agency announced the preliminary results of its evaluation in May of 1999, focusing on the following criteria: (i) whether environmental or human health improvements result from the policy; (ii) whether the policy encourages prompt disclosure and correction of violations; (iii) whether the policy encourages improvements in corporate compliance programs; and (iv) whether the policy engenders greater awareness of new environmental issues. EPA found in conducting its evaluation that:

- Discovery and correction of violations removed pollutants from the environment, reduced health and environmental risks, and improved public information on potential hazards.
- The policy has been applied consistently.
- In 27 instances, new environmental issues came to the Agency's attention as a result of application of the policy.
- Use of the policy has been widespread.[29]
- The rate of disclosure has increased every year since the policy has been in place.
- Users had a high satisfaction rate: 88 percent said they would use it again and 84 percent said they would recommend others to use the policy.
- Most disclosures involved monitoring and reporting violations that occurred in federally run programs, i.e., programs other than ones that states are authorized, delegated, or approved to administer and enforce.
- The policy encourages improvements in auditing programs and environmental management systems.
- The most commonly suggested changes were to

at 5; *EPA Gets Broad Support on Final Audit Policy, Though Concerns Linger*, INSIDE EPA WKLY. REP., Dec. 22, 1995, at 11; Paul J. Curran & Gregory J. Wallace, *The New EPA "Interim Policy" Which Is Meant to Encourage Companies to Report Violations, May Have the Opposite Effect*, NAT'L L.J., July 31, 1995, at B4; Steven A. Herman, *EPA's New Voluntary Environmental Self-Policing and Self-Disclosure Interim Policy Statement*, NAT'L ENVTL. ENFORCEMENT J., Apr. 1995, at 15; *EPA Releases Explanatory Document on Audit Policy, Self-Disclosure Cases*, Daily Rep. for Executives (BNA), Apr. 19, 1995, at A-13; *Many Violations May Not Qualify for Penalty Reductions Under Audit Policy*, Daily Rep. for Executives (BNA), Apr. 19, 1995, at A-31; *EPA: Confess and Pay Less*, NAT'L L.J., Apr. 17, 1995, at A-6; *New Audit Policy Creates Uncertainty Obstacles to Audits, Industry Counsel Says*, Daily Rep. for Executives (BNA), Apr. 4, 1995, at A-21.

25. Under the 1995 Self-Policing Policy, prompt disclosure was defined as within 10 days after discovery of the violation. 1995 Self-Policing Policy, *supra* note 22, at 66711.

26. Memorandum from Steven Herman, Assistant Administrator, EPA, to Regional Administrators and Assistant Attorney General, Environment and Natural Resources Division (Jan. 15, 1997) (available from the ELR Document Service, ELR Order No. AD-3124).

27. The Audit Policy Update is available at <http://www.epa.gov/oeca/apolguid.html>.

28. U.S. EPA, Agency Information Collection Activities: Proposed Collection; Comment Request; Audit Policy Customer Satisfaction Survey, 63 Fed. Reg. 25855 (May 11, 1998). Participation in the survey was voluntary and respondents could remain anonymous if they so wished.

29. As of the close of 1999, approximately 500 entities had disclosed violations at over 2,700 facilities.

expand the 10-day disclosure period and to shorten the amount of time taken to process cases.[30]

EPA's 2000 Self-Policing Policy

On April 11, 2000, EPA promulgated relatively minor revisions to its 1995 Self-Policing Policy.[31] Effective May 11, 2000, the revisions maintain the basic structure and terms of the 1995 Policy, while clarifying some of its language, broadening its applicability, and conforming the provisions of the policy to actual agency practice. The revisions lengthen the period for prompt disclosure from 10 to 21 days,[32] clarify that the independent discovery period does not automatically preclude penalty mitigation for multi-facility entities,[33] and detail how the prompt disclosure and repeat violation conditions apply to newly acquired companies.[34] The revised policy also clarifies the independent discovery condition by noting that when EPA determines that the facility did not know it was under civil investigation and finds that the entity is otherwise acting in good faith, the Agency may exercise its discretion to reduce or waive civil or administrative penalties.[35] Unlike the 1995 Policy, which required the disclosing entity to meet all nine conditions of the policy in order to qualify for an incentive in a criminal context, the revisions provide that EPA will generally elect not to recommend criminal prosecution by the DOJ or other prosecuting authority for a disclosing entity that meets at least conditions two through nine—regardless of whether it meets the systematic discovery requirement — as long as its self-policing, discovery and disclosure were conducted in good faith and the entity adopts a systematic approach to preventing recurrence of the violation.[36]

The 2000 Self-Policing Policy uses the term "compliance management system" instead of "due diligence," which was used in the 1995 Policy.[37] This change in nomenclature is intended solely to conform the 2000 Policy to terminology more commonly used by industry and by regulators to refer to a systematic management plan or systematic efforts to achieve and maintain compliance with environmental laws. Intending no substantive difference in the change, the 2000 Policy clearly indicates that a compliance management system must reflect the regulated entity's due diligence in preventing, detecting, and correction violations.[38] The 2000 Policy also states that in some instances, certain Clean Air Act violations discovered, disclosed, and corrected by a company prior to issuance of a Title V permit are eligible for penalty mitigation under the policy.[39]

In addition, the 2000 Self-Policing Policy clarifies somewhat how the disclosing entity must cooperate with EPA in order to meet the requirements of the ninth condition of the policy. The entity must cooperate as required by EPA and provide such information as is necessary and required by EPA to determine the applicability of the policy to the entity.[40] The entity must not hide, destroy, or tamper with possible evidence of potential environmental violations. In most instances, EPA will not request company audit reports.[41] The revised policy further clarifies that entities who disclose potential criminal violations may expect a more thorough review by the agency than in cases of disclosures of civil violations:

> In criminal cases, entities will be expected to provide, at a minimum, the following: access to all requested documents; access to all employees of the disclosing entity; assistance in investigating the violations, any noncompliance problems related to the disclosure, and any environmental consequences related to the violations; access to all information relevant to the violations disclosed, includ-

30. Nancy K. Stoner & Amanda A. Gibson survey the benefits of the 1995 Self-Policing Policy in *EPA's Audit Policy Spells Success for Corporate Users, EPA, the Public, and Most Importantly, the Environment*, 29 ELR 10705 (Nov. 1999).

31. U.S. EPA, Incentives for Self-Policing, Discovery, Correction and Prevention of Violations, 65 Fed. Reg. 19618 (Apr. 11, 2000) [hereinafter 2000 Self-Policing Policy]. Technically, the 2000 Self-Policing Policy supercedes the 1995 Self-Policing Policy. *Id.*, at 19618, 19626. For commentary on EPA's proposed revisions to the 1995 Self-Policing Policy, *see* U.S. EPA, Evaluation of "Incentives for Self-Policing: Discovery, Disclosure, Detection and Prevention of Violations" Policy Statement, Proposed Revisions and Request for Public Comments, 64 Fed. Reg. 26745 (May 17, 1999) (available from the ELR Document Service, ELR Order No. AD-4143). *See generally* Stoner & Gibson, *supra* note 30, at 10707; Cheryl Hogue, *Proposed Changes to EPA Policy Would Extend Deadline for Disclosure to 21 Days*, Daily Env't Rep. (BNA), May 14, 1999, at AA-1. On the Agency's reevaluation of the 1995 Self-Policing Policy, *see* Elizabeth Glass Geltman, *EPA's Environmental Audit Policy Turns Five!*, ENVTL. TRANSACTIONS, AUDITS & BROWNFIELDS COMMITTEE NEWSL. (American Bar Ass'n), Dec. 1999, at 2.

32. 2000 Self-Policing Policy, *supra* note 31, at 19618, 19621. The 21-day disclosure period begins when the entity discovers that a violation has, or may have, occurred. The trigger for discovery is when any officer, director, employee or agent of the facility has an "objectively reasonable basis" for believing that a violation has, or may have, occurred. The 2000 Policy's "objectively reasonable basis" standard is measured against what a prudent person, having the same information as was available to the individual in question, thought was reasonable at the time the situation was encountered. If an entity has some doubt as to the existence of a violation, the recommended course is for the entity to proceed with the disclosure and allow the regulatory authorities to make a definitive determination. Contract personnel who provide on-site services at the facility may be treated as employees or agents for purposes of the 2000 Policy. *Id.* at 19621.
The 2000 Policy clarifies how to calculate the 21-day period if the 21st day falls on a weekend or federal holiday. For situations in which the 21-day period has already expired, the 2000 Policy expressly allows that EPA may accept a late disclosure in an exceptional case, such as where there are complex circumstances, including where EPA determines that the violations could not be identified and disclosed within 21 calendar days after discovery. EPA may also extend the 21-day disclosure period when multiple facilities or acquisitions are involved. *Id.* at 19621-22.

33. *Id.* at 19618, 19622, 19626. EPA encourages multi-facility auditing and does not intend for the "independent discovery" condition to preclude the availability of the Self-Policing Policy when multiple facilities are involved. Accordingly, if a regulated entity owns or operates multiple facilities, the fact that one of them is the subject of any investigation, inspection, information request or third-party complaint does not automatically preclude EPA from granting Self-Policing Policy credit for the disclosures of violations self-discovered at the other facilities, assuming all other conditions of the policy are met. *Id.* at 19622, 19626.

34. *Id.* at 19622-23.

35. *Id.* at 19622, 19626.

36. *Id.* at 19620, 19624.

37. *Id.* at 19625.

38. *Id.* at 19621.

39. *Id.*; for further guidance *see* Memorandum from Eric Schaeffer, Director, Office of Regulatory Enforcement, U.S. EPA, to Steven A. Herman, Assistant Administrator et al., on Reduced Penalties for Disclosures of Certain Clean Air Action Violations (Sept. 30, 1999). The document is available on the Internet at <www.epa.gov/oeca/ore/apolguid.htm>.

40. 2000 Self-Policing Policy, *supra* note 31, at 19623, 19626.

41. *Id.* at 19623.

ing that portion of the environmental audit report or documentation from the compliance management system that revealed the violation; and access to the individuals who conducted the audit or review.[42]

Lastly, the 2000 Self-Policing Policy recognizes that where states have adopted their own audit and self-policing policies that pertain to federally authorized, approved, or delegated programs, EPA will generally defer to state penalty mitigation for self-disclosures as long as the state policy meets minimum requirements for federal authorization, approval, or delegation.[43]

Shielding Materials From Disclosure Under Common-Law Doctrines and State Audit Laws

Seneca's maxim: "If you would wish another to keep your secret, first keep it to yourself,"[44] bears keeping in mind when contemplating an environmental audit. Neither traditional common-law doctrines such as the attorney client privilege and the work product doctrine, nor nascent rubrics such as the "privilege" of self-critical analysis will avail much in preventing the disclosure of materials prepared by auditors.[45] Particularly for firms in the United States, the auditing process involves the creation of materials that could later be used against the company in an enforcement contest or litigation with a third party.[46] Legislators in several states have enacted laws shielding environmental audit materials from disclosure in certain situations in answer to such concerns, while lawmakers in other jurisdictions have not, reasoning that such insulation is inappropriate.[47] [Appendix A contains a summary of the state audit privilege and immunity statutes and self-disclosure policies.] Nonetheless, even if audit documents are protected from disclosure under a state statute, the document may still have to be disclosed to EPA or to litigants in another jurisdiction. A consensus appears to be growing, however, in favor of a level of protection against criminal charges for firms that have genuine comprehensive audit systems, where prompt response and correction are made to any environmental vio-

42. *Id.*; *see also* Memorandum from Earl E. Devaney, Director, Office of Criminal Enforcement, U.S. EPA, to All EPA Employees Working in or in Support of the Criminal Enforcement Program, Regarding Guidance on EPA's Exercise of Investigative Discretion for Environmental Crimes (Jan. 12, 1994) (available from the ELR Document Service, ELR Order No. AD-3418); U.S. Department of Justice "Factors" Policy, *supra* note 21.

43. 2000 Self-Policing Policy, *supra* note 31, at 19624; *see also* Memorandum from Steven A. Herman, Assistant Administrator, Office of Enforcement and Compliance Assurance, U.S. EPA, et al., to Regional Administrators, regarding Statement of Principles: Effect of State Audit Immunity/Privilege Laws on Enforcement Authority for Federal Programs (Feb. 14, 1997).

44. SENECA, TRAGEDIES 725, 757 (Harvard Univ. Press, Loeb Classical Series 1917) (Frank J. Miller trans.).

45. *See* Keith Casto & Tiffany Billingsley Potter, *Environmental Audits: Barriers, Opportunities and a Recommendation*, 5 HASTINGS W.-NW J. ENVTL. L. & POL'Y 233, 245-49 (1999); E. Lynn Grayson & Christina M. Riewer, *EPA's Audit Policy and State Audit-Privilege Laws: Moving Beyond Command and Control?*, 27 ELR 10243, 10249-50 (May 1997) [hereinafter Grayson & Riewer]; Donald A. Carr & William L. Thomas, *Devising a Compliance Strategy Under the ISO 14000 International Environmental Management Standards*, 15 PACE ENVTL. L. REV. 85, 193-205 (1997) [hereinafter Carr & Thomas].

46. *See generally* Philip D. Reed, *Environmental Audits and Confidentiality: Can What You Know Hurt You as Much as What You Don't Know?*, 13 ELR 10303 (Oct. 1983); *see also* CORPORATE COUNSEL'S GUIDE TO ENVIRONMENTAL COMPLIANCE AND AUDITS (Business Laws, Inc. 1992 and 1998 Supp.).

47. In July 1993, Oregon became the first state in the United States to codify a self-evaluative privilege covering environmental audits. The new privilege bears some similarity to provisions in the *Federal Rules of Evidence* and many state evidence codes that prohibit the use of evidence of subsequent remedial measures to establish a party's negligence. As of July 2000, 24 states have adopted some form of environmental audit privilege and/or penalty immunity law. *See, e.g.*, Privilege and Immunity: Colorado (COLO. REV. STAT. §13-25-126.5, 25-1-114.5 (1994)), Kentucky (KY. REV. STAT. ANN. §224.01-040 (Michie 1994)), Minnesota (MINN. STAT. ANN. §114C.20 et seq. (West 1995)), Wyoming (WYO. STAT. ANN. §35-11-1105 et seq. (Michie 1995)), Utah (UTAH CODE ANN. §19-7-101 et seq. (1995)), Texas (TEXAS REV. CIV. STAT. ANN. art. 4447cc (West 1995)), Kansas (KAN. STAT. ANN. §60-3332 et seq. (1995)), Virginia (VA. CODE ANN. §10.1-1198 et seq. (Michie 1995)), Michigan (MICH. COMP. LAWS ANN. §324.14801 et seq. (1996)), South Carolina (S.C. CODE ANN. §48-57-10 et seq. (Law. Co-op. 1996)), New Hampshire (N.H. REV. STAT. ANN. §147-E:1 et seq. (1996)), Ohio (OHIO REV. CODE ANN. §3745.70 et seq. (Anderson 1996)), Montana (MONT. CODE ANN. §75-1-1202 et seq., 80-1-301 et seq. (1997)), Alaska (ALASKA STAT. §9.25.450 et seq. (Michie 1997)), Nevada (NEV. REV. STAT. §445C.010 et seq. (1997)), Nebraska (NEB. REV. STAT. §25-21, 254 et seq. (1998)), Iowa (IOWA CODE §1109 (1998)), South Dakota (S.D. CODIFIED LAWS §1-40-33 et seq. (Michie 1996)); Privilege Only: Oregon (ORE. REV. STAT. §468.963 (1993)), Indiana (IND. CODE §13-28-4-1 et seq. (1994)), Illinois (ILL. COMP. STAT. ANN. §415-5/52.2 (West 1995)), Arkansas (ARK. CODE ANN. §8-1-301 et seq. (Michie 1995)), Mississippi (MISS. CODE ANN. §49-2-71 (1995)); Immunity Only: Rhode Island (R.I. GEN. LAWS §42-17.8-1 et seq. (1997)). Idaho also enacted an audit privilege/immunity law in 1995, but it sunset on December 31, 1997 (IDAHO CODE §9-801 et seq. (1995)). In April 2000, Arizona enacted a law that contains elements of environmental audit privilege and immunity in the context of encouraging environmental management systems (ARIZ. REV. STAT. tit. 49-171 et seq. (2000)). Since 1994, EPA has devoted considerable time and effort trying to persuade states to amend their audit privilege and/or immunity laws where enactment has in EPA's view created legal insufficiencies as to the minimum state authorities necessary to carry out federally authorized, delegated, or approved state programs. As of July 2000, 14 of 24 states with such laws have enacted statutory amendments and/or issued a clarifying Attorney General's opinion to satisfy EPA's legal concerns with the federally authorized, delegated, or approved state programs in these states. *See* Appendix A. Audit privilege laws generally protect environmental information related to an audit from disclosure to the public or regulatory agencies and prohibit the use of the information in judicial or administrative proceedings, including enforcement settlements. *See* Nancy K. Stoner & Wendy J. Miller, *National Conference of State Legislatures Study Finds That State Environmental Audit Laws Have No Impact on Company Self-Auditing and Disclosure of Violations*, 29 ELR 10265 (May 1999) [hereinafter Stoner & Miller]. Audit immunity laws typically provide immunity from penalties for violations discovered during an audit and reported to a regulatory agency. *See* Stoner & Miller, at 10265-66.

Proponents of audit privilege and/or immunity laws suggest that the laws are needed to encourage companies that do not currently audit to begin to do so; to encourage more auditing by companies that already audit; and for immunity laws, to encourage disclosure and prompt correction of environmental violations. *See id.* at 10266. EPA has consistently opposed such laws because of their effect on public access to information and on the states' ability to enforce regulations to protect human health and the environment, as explained in, among other sources, Stoner & Miller, *id.* at 10266 n.5; Lois J. Schiffer & Ann C. Juliano, *Reform of Environmental Regulations: Three Points*, NAT. RESOURCES & ENV'T, Winter 1998, at 175; Nancy Stoner, The Future of Audit Privilege, Presentation at the American Bar Association, Section of Natural Resources, Energy, and Environmental Law Fall Meeting, Asheville, N.C. (Oct. 1997); *Official Says GOP Bills Will Deter EPA From Tracking Evidence of Criminal Conduct*, Daily Rep. for Executives (BNA), Sept. 11, 1997, at A-17. Just as it has fought federal legislation, EPA has resisted state audit privilege, immunity, and amnesty-related legislation. Memorandum from Nancy K. Stoner, Director, Office of Planning & Policy Analysis, EPA, to Regional Administrators et al., regarding the Subject of State Audit Privilege/Immunity Laws and

lations that are uncovered.[48] Yet the balance between providing this incentive for improved compliance, and preventing the manipulative use of "audits of convenience" as cover for violations, has and will continue to be difficult to define and implement.

The limited assurances and penalty mitigation of the kind offered under EPA's and the DOJ's policies described above[49] have had a positive impact but have not eliminated industry fears altogether.[50] A 1998 study by the National Conference of State Legislatures[51] found that although the use of audits in enforcement remains a significant concern among members of the regulated community, it has not affected auditing rates.[52] Other previous studies by industry have found that some members of the regulated community have been reluctant to perform audits for fear of the audit being later used in an enforcement case or litigation with third parties, but that most mid-to-large sized company respondents performed audits for business reasons.[53] As the risk that disclosure will come back to haunt the auditing company in a variety of realistic litigation scenarios remains, many firms will continue to struggle with audit risk/benefit analysis. In recent years, EPA and the DOJ have sought to increase the number of regulated entities performing audits by requiring commitments to such self-evaluations in a number of consent decrees and administrative orders. [Appendix B summarizes the audit provisions contained in federal court consent decrees or orders; Appendix C summarizes those found in EPA administrative orders.]

Auditing as a Factor in Enforcement Settlements, Criminal Sanctions, and Delisting

For the most part, neither EPA's general civil penalty policy[54] nor the various program-specific civil penalty policies[55] directly address the issue of environmental auditing. Prior to the 1995 Self-Policing Policy, where consideration of environmental auditing in penalty mitigation has occurred under a program-specific penalty policy, it has almost always been on a case-by-case basis.[56] One exception

Legislation (Nov. 10, 1997) ("EPA opposes such laws because they limit access to evidence of civil violations or criminal misconduct, thereby complicating criminal and civil discovery and frustrating public access to information."); Memorandum from Steven A. Herman et al., EPA, to Regional Administrators, regarding the subject of the Effect of State Audit Immunity/Privilege Laws on Enforcement Authority for Federal Programs (Feb. 14, 1997). State audit privilege and immunity laws have been analyzed by a number of commentators. *E.g.*, <http://www.epa.gov/region5/audits/audit_articles.htm> (offering access to many articles, and updated annually); Timothy T. Jones et al., *Environmental Compliance Audits: The Arkansas Experience*, 21 U. ARK. LITTLE ROCK L. REV. 191 (1999); Bertram C. Frey & Matthew J. Sanders, Cooperation With Government Agencies Under Environmental Audit Laws, Rules and Policies, ABA Section of Natural Resources, Energy, and Environmental Law, 27th Annual Conference on Environmental Law, Keystone, Colo. (Mar. 1998); John-Mark Stensvaag, *The Fine Print of State Environmental Audit Privileges*, 16 UCLA J. ENVTL. L. & POL'Y 69 (1997-1998); Michael S. Mostek, *Limited Privilege and Immunity for Self-Evaluative Environmental Audits in Nebraska: Moving Environmental Performance to the Next Level*, 32 CREIGHTON L. REV. 545 (1998); Patrick Decker Sachse, *Environmental Audits: An Analysis of the Dilemma and an Assessment of Oklahoma's Response*, 51 OKLA. L. REV. 541 (1998); Joan M. Fildes, *Statutory Privileges and Immunities for Voluntarily Performed Environmental Audits: Should New York State Join the Race?*, 5 BUFF. ENVTL. L.J. 257 (1998); John A. Lee & Bertram C. Frey, *Environmental Audit Immunity Laws: A State-by-State Comparison*, 28 Env't Rep. (BNA) 331 (June 13, 1997); Lynn Holdsworth, *Florida's Environmental Self-Audit Legislation: An Incentive for the Environmentally Conscientious Business or an Opportunity for the Corporate Polluter to Suppress the Truth?*, 27 STETSON L. REV. 210 (1997); Elizabeth Glass Geltman & Carey Ann Mathews, *Environmental Democracy*, 22 J. CORP. L. 395 (1997); Grayson & Riewer, *supra* note 45, at 10246-49; Anna Kathryne Campbell, *The Environmental Audit Privilege: Where Does Louisiana Stand in the Federal v. State Showdown?*, 57 LA. L. REV. 1029 (1997); Susan J. Spicer, *Turning Environmental Litigation on Its E.A.R.: The Effects of Recent State Initiatives Encouraging Environmental Audits*, 8 VILL. ENVTL. L.J. 1 (1997); Christina Austin, *State Environmental Audit Privilege Laws: Can EPA Still Access Environmental Audits in Federal Court?*, 26 ENVTL. L. 1241 (1996); Clinton J. Elliott, *Kentucky's Environmental Self-Audit Privilege: State Protection or Increased Federal Scrutiny?*, 23 N. KY. L. REV. 1 (1995); Edward Felsenthal, *New Laws Make Companies' Self-Reviews Less Accessible*, WALL ST. J., Aug. 2, 1994, at B2; James T. O'Reilly, *Environmental Audit Privileges: The Need for Legislative Recognition*, 19 SETON HALL L. REV. 119 (1994).

48. *See* Michael Herz, *Environmental Auditing and Environmental Management: The Implicit and Explicit Federal Regulatory Mandate*, 12 CARDOZO L. REV. 1241 (1992).

49. Seventeen states have also adopted some form of self-disclosure policy. They are: California (July 8, 1996, revised Dec. 1998); Connecticut (Oct. 30, 1996, revised Mar. 27, 2000); Delaware (Apr. 11, 1997); Florida (Apr. 1, 1996); Indiana (Apr. 5, 1999); Maine (Jan. 1, 1996, revised Feb. 14, 1996) (only applicable to small businesses); Maryland (June 24, 1997); Massachusetts (Apr. 26, 1997); Minnesota (Jan. 24, 1995); New Mexico (Feb. 5, 1999); New York (Aug. 12, 1999) (only applicable to small businesses); North Carolina (Sept. 1, 1995); Oregon (Feb. 18, 1999); Pennsylvania (Sept. 25, 1996); Tennessee (Nov. 27, 1996); Vermont (Dec. 14, 1996, sunset Dec. 1997); and Washington (Dec. 20, 1994). *See* Appendix A.

50. *See* Bergeson & Campbell, *supra* note 24, at 38, 41 ("In a nutshell industry does not think the audit policy goes far enough and many state governments, as well as members of Congress agree."); Brooks M. Beard, *The New Environmental Federalism: Can the EPA's Voluntary Audit Policy Survive?*, 17 VA. ENVTL. L.J. 1 (1997); Daniel M. Steinway, *EPA's Auditing Policy: Is It Working?*, METROPOLITAN CORP. COUNS., May 1997, at 15; Lawrence S. Kirsch & J. Walter Veirs, *Although the EPA Recently Issued Its Final Audit Policy on the Confidentiality of a Company's Environmental Self-Audit, Numerous Questions Remain Unanswered*, NAT'L L.J., Mar. 11, 1996, at B5, B6 (companies "that voluntarily disclose violations continue to risk that the EPA and state agencies, as well as private litigants, will use the information disclosed against them.").

51. LARRY MORANDI, STATE ENVIRONMENTAL AUDIT LAWS AND POLICIES: AN EVALUATION (National Conference of State Legislatures 1998).

52. *See* Stoner & Miller, *supra* note 47, at 10268. A survey of 369 companies performed in 1995 by Price Waterhouse revealed that 75 percent of respondents had established a program to audit regulatory compliance. *See* PRICE WATERHOUSE LLP, THE VOLUNTARY ENVIRONMENTAL AUDIT SURVEY OF U.S. BUSINESS (1995). The results of the Price Waterhouse survey are consistent with those of another poll in which nearly all of the 78 respondents reported that they regularly perform regulatory compliance audits. *See* COOPERS & LYBRAND, STRATEGIC ENVIRONMENTAL MANAGEMENT PREPAREDNESS FOR ISO 14000—A Survey (1995). Of this number, slightly more than one-half had also audited their environmental management systems. *Id.* According to an industry survey, the number of firms performing audits would increase, however, should audit information be given protection. *See New Survey Shows Enforcement Questions Block Use of Voluntary Audits*, INSIDE EPA WKLY. REP., Apr. 14, 1995, at 4; *Companies Would Perform More Audits if Penalties Were Eliminated, Survey Says*, 25 Env't Rep. (BNA) 1606 (Apr. 14, 1995).

53. *See id.*

54. Memorandum from Courtney M. Price, Assistant Administrator for Enforcement and Compliance Monitoring, EPA, to Associate Administrators, Assistant Administrators, General Counsels, Inspector General, Regional Administrators and Staff Office Directors on New Civil Penalty Policy (Feb. 16, 1984).

55. *E.g.*, A FRAMEWORK FOR STATUTE-SPECIFIC APPROACHES TO PENALTY ASSESSMENTS: IMPLEMENTING EPA'S POLICY ON CIVIL PENALTIES (1984).

56. *See* discussion of EPA's supplemental environmental projects policy *infra* at note 63.

is the penalty policy under the Toxic Substances Control Act (TSCA), which considers self-confession and environmental auditing in penalty mitigation.[57]

In the area of sentencing, EPA's July 24, 1992, guidance entitled *Procedures to Implement the Guidelines of the U. S. Sentencing Commission for Organization Defendants*[58] makes clear that environmental auditing can be imposed as a condition of probation. The sentencing guidelines themselves, which expand the use of remedial orders as conditions of probation, contain explicit provisions for the imposition of measures that can be characterized as auditing.[59] Moreover, courts can consider the existence of an effectively functioning environmental auditing program when determining an appropriate sentence.[60]

Environmental auditing also bears on the collateral consequences of violations. EPA can choose to suspend or debar any party that violates, or is suspected of violating, environmental laws from federal contracting. EPA's October 31, 1991, policy regarding contractor suspension, debarment and delisting requires, as a practical matter, corporations desirous of being removed from the Agency's list of ineligible facilities to create and implement an effective compliance audit program.[61] Such a program must include a commitment to expeditiously correct any violations once they have been identified.[62]

Most significant for purposes of this Monograph is EPA's February 1991, *Policy on the Use of Supplemental Environmental Projects in EPA Settlements*, updated effective May 1, 1998.[63] The Agency's 1998 policy explicitly recognizes that certain types of environmental assessments and audits will qualify to reduce the gravity amount of a civil penalty, including pollution prevention assessments, environmental quality assessments, and compliance audits. The defendant or respondent must agree to provide EPA with a copy of the audit report and usually must also commit that after conducting the audit it will correct any deficiencies, both with respect to management and environmental practices, uncovered by the audit. In general, compliance audits are acceptable as supplemental environmental projects (SEPs) only when the defendant or respondent is a small business or small community.[64]

To date, the DOJ and EPA have negotiated over 150 consent decrees requiring environmental auditing. While some consent decrees merely mention auditing in general terms,[65] many are quite specific.[66] Generally, such decrees and orders delineate the type of audit to be performed (e.g., field audit, lab audit, facility audit, and/or compliance audit) and indicate whether or not the entity must file reports as part of the audit.[67] Almost all the decrees also require some form of reporting of audit results to governmental authorities. In the case of orders and decrees arising under the Comprehensive Environmental Response, Compensation, and Liability Act (CERCLA or, as it is often referred, Superfund), it is significant that almost all require audits of laboratories that analyze samples of hazardous substances, as well as field audits of sampling procedures and techniques at Superfund sites. Orders and decrees under EPA's other programs typically require compliance audits at specific facilities. A few settlements require that the company pay stipulated penalties for redress of violations uncovered in any audit.[68] The frequency of the audit(s) is sometimes specified, but more often is left to the entity's discretion.

Reducing Pollution, and Penalties, With SEPs

It is the national policy of the United States that:

> pollution be prevented or reduced at the source whenever feasible; pollution that cannot be prevented should be recycled in an environmentally safe manner, whenever feasible; pollution that cannot be prevented or recycled should be treated in an environmentally safe manner whenever feasible; and disposal or release into the environment should be employed only as a last resort and should be conducted in an environmentally safe manner.[69]

57. U.S. EPA, Office of Enforcement, Guidelines for the Assessment of Civil Penalties Under Section 16 of the Toxic Substances Control Act and PCB Penalty Policy, 45 Fed. Reg. 59770 (Sept. 10, 1980) (available from the ELR Document Service, ELR Order No. AD-519).

58. Memorandum from Herbert H. Tate Jr., Assistant Administrator for Enforcement, EPA, to Regional Administrators, Deputy Regional Administrators, and Regional Counsels (July 24, 1992).

59. U.S. Sentencing Commission Guidelines, §8D1.4(c). The guidelines can be accessed at <http://www.ussc.gov/>.

60. Memorandum from Herbert H. Tate Jr., *supra* note 58 at 5.

61. U.S. EPA, EPA Policies Regarding the Role of Corporate Attitude, Policies, Practices, and Procedures in Determining Whether to Remove a Facility From the EPA List of Violating Facilities Following a Criminal Conviction, 56 Fed. Reg. 64785 (Dec. 12, 1991), ADMIN. MAT. 35411 (available from the ELR Document Service, ELR Order No. AD-3254). *See generally* Carol E. Dinkins, *Collateral Consequences of Conviction*, *in* ENVIRONMENTAL CRIMINAL LIABILITY: AVOIDING AND DEFENDING ENFORCEMENT ACTIONS 318 (Donald A. Carr ed., 1995).

62. 56 Fed. Reg. at 64789.

63. Memorandum from James M. Strock, Assistant Administrator, EPA, to Regional Administrators, Deputy Regional Administrators, Regional Counsels, Regional Program Division Directors, Assistant Administrators, General Counsels, Program Compliance Directors and Associate Enforcement Counsels (Feb. 12, 1991), ADMIN. MAT. 35607 (available from the ELR Document Service, ELR Order No. AD-123). The SEP policy was most recently updated effective May 1, 1998. *See* Memorandum from Steven A. Herman, Assistant Administrator, EPA, to Regional Administrators regarding Issuance of Final Supplemental Environmental Projects Policy (Apr. 10, 1998), ADMIN. MAT. 35703 (available from the ELR Document Service, ELR Order No. AD-3749) [hereinafter 1998 SEP Policy].

64. 1998 SEP Policy, *supra* note 63, at D.5. This is a significant change from the previous SEP policy, which most likely reflects the fact that most large companies are already conducting internal audits.

65. More than 50 other federal consent decrees refer to environmental auditing, but only in the most general of terms, and therefore are not discussed here.

66. EPA has also entered into a number of informal agreements wherein companies have agreed to perform audits, *see, e.g.*, Arizona Chemical Company (December 1998), but such commitments have not been incorporated in a judicial or administrative order.

67. Fifteen consent decrees also require an audit of a company's EMS.

68. *E.g.*, United States v. Darling Int'l, Inc., No. C.V. 97-1611 (D. Minn., order entered Sept. 12, 1997); United States v. Rhone-Poulenc Agric. Co., No. C.V. 2:90-0929 (S.D. W. Va., order entered Dec. 19, 1991); United States v. Wheeling-Pittsburgh Steel Corp. (Allenport Facility), No. C.V. 89-237 S (W.D. Pa., order entered Nov. 4, 1992). Other consent decrees require the payment of stipulated penalties for the violation of auditing provisions or of the audit workplan, but not for the violations themselves discovered under such provisions or workplan. *E.g.*, United States v. E.I. Dupont de Nemours & Co. (Chamber Works), No. C.V. 91-768 (D.N.J., order entered May 21, 1991); United States v. Grumman St. Augustine Corp., No. C.V. 91-141-CIV-J-16 (M.D. Fla., order entered July 20, 1993); United States v. New Haven Foundry, Inc., No. C.V. 97-71842 (E.D. Mich., order entered June 23, 1997).

69. The Pollution Prevention Act of 1990, 42 U.S.C. §13101, ELR STAT. PPA §3. *See generally* CYNTHIA A. LEWIS & JAMES M. THUNDER,

EPA's goal, as set forth in the pollution prevention strategy announced in 1991, is to "institutionalize" the pollution prevention ethic.[70] Source reduction is gradually taking hold, primarily as a result of a growing appreciation within industry of the implications of pollution prevention for firm profitability.[71]

While much of its activity in the pollution prevention area is aimed at promoting voluntary action,[72] the Agency has not shied away from using its leverage in enforcement negotiations to encourage firms to adopt such practices.[73] In many cases, pollution prevention programs are required to redress violations and achieve and maintain compliance with environmental regulations, often as specific conditions of settlements. With increasing regularity, companies are mitigating cash fines and penalties by undertaking source reduction and similar techniques as SEPs.[74] [Appendix D summarizes the pollution prevention SEPs entered into in fiscal year 1999.]

A SEP is an "environmentally beneficial project" that is

FEDERAL CHEMICAL REGULATION: TSCA, EPCRA AND THE POLLUTION PREVENTION ACT (1997); E. Lynn Grayson, *The Pollution Prevention Act of 1990: Emergence of a New Environmental Policy*, 22 ELR 10392 (June 1992). On pollution prevention at the state level, *see* Michele Ochsner, *Pollution Prevention: An Overview of Regulatory Incentives and Barriers*, 6 N.Y.U. ENVTL. L.J. 586, 617 n. 57 (1998) (citing sources) [hereinafter Ochsner]; Robert F. Blomquist, *Government's Role Regarding Industrial Pollution Prevention in the United States*, 29 GA. L. REV. 349 (1995) [hereinafter Blomquist].

70. U.S. EPA, Pollution Prevention Strategy, 56 Fed. Reg. 7849 (Feb. 26, 1991). As Prof. Strasser observed, EPA's definition of pollution prevention, which emphasizes a policy preference for source reduction over other waste reuse, treatment, or disposal methods, has given rise to "controversy and confusion." Kurt A Strasser, *Preventing Pollution*, 8 FORDHAM ENVTL. L.J. 1, 5 n.10 (1996) [hereinafter *Preventing Pollution*].

71. *See generally* Richard Florida, *Research Suggests EMSs, P2 Programs Favor Improved Performance*, Int'l Envtl. Sys. Update, Dec. 1999, at 14; Forest L. Reinhardt, *Bringing the Environment Down to Earth*, HARV. BUS. REV., July/Aug. 1999, at 149; Forest L. Reinhardt, *Market Failure and the Environmental Policies of Firms*, 3 J. INDUS. ECOLOGY 9 (1999); PAUL HAWKEN ET AL., NATURAL CAPITALISM (Little, Brown & Co. 1999); Amory Lovins et al., *A Road Map for Natural Capitalism*, HARV. BUS. REV., May/June 1999, at 145; Daniel C. Esty & Michael E. Porter, *Industrial Ecology and Competitiveness: Strategic Implications for the Firm*, 2 J. INDUS. ECOLOGY 35 (1998); JUDITH A. CICHOWITZ, HOW TO CONTROL COSTS IN YOUR POLLUTION PREVENTION PROGRAM (John Wiley & Sons 1997); JAMES R. ALDRICH, POLLUTION PREVENTION ECONOMICS: FINANCIAL IMPACTS ON BUSINESS AND INDUSTRY (McGraw Hill 1995); Kenneth E. Nelson, *Finding and Implementing Projects That Reduce Waste*, *in* INDUSTRIAL ECOLOGY AND GLOBAL CHANGE 371 (Robert Socolow et al. eds., Cambridge University Press 1994); MARK H. DORFMAN ET AL., ENVIRONMENTAL DIVIDENDS: CUTTING MORE CHEMICAL WASTES (1992). Some commentators are skeptical. *See, e.g.*, Ochsner, *supra* note 69, at 616 n.2 (citing sources); *When Pollution Meets the Bottom Line*, ENVTL. SCI. & TECH., Sept. 1997, at 418A. The periodical *P2: Pollution Prevention Review* features case studies of firms from the United States and around the world. Other success stories can be found on EPA's website at <http://www.epa.gov/opptintr/3350>. Another repository of P2 knowledge is the National Pollution Prevention Roundtable. *See* <http://www.p2.org/>.

72. EPA's Office of Pollution Prevention and Toxics sponsors a number of programs and activities that are intended to encourage the regulated community to adopt pollution prevention programs. Many of these initiatives are described on the Agency's website at <http://www.epa.gov/opptintr/p2home/programs.htm>. For additional background, *see* NATIONAL ADVISORY COUNCIL FOR ENVIRONMENTAL POLICY AND TECHNOLOGY, U.S. EPA, TRANSFORMING ENVIRONMENTAL PERMITTING AND COMPLIANCE POLICIES TO PROMOTE POLLUTION PREVENTION, EPA/100-R-93-004 (1993); OFFICE OF RESEARCH AND DEVELOPMENT, U.S. EPA, FACILITY POLLUTION PREVENTION GUIDE, EPA/600-R-92-088 (1992); *see generally* Michael Zatza & Shana Harhourb, *The United States Environmental Protection Agency's 33/50 Program: The Anatomy of a Successful Voluntary Pollution Reduction Program*, J. CLEANER PRODUCTION, Feb. 1999, at 17; Ochsner, *supra* note 69; Kurt A. Strasser, *Cleaner Technology, Pollution Prevention and Environmental Regulation*, 9 FORDHAM ENVTL. L.J. 1 (1997) [hereinafter Strasser]; *Preventing Pollution*, *supra* note 70; Blomquist, *supra* note 69; Joel S. Hirschorn, *Pollution Prevention Comes of Age*, 29 GA. L. REV. 325 (1995); Carol Foley & Michael Elliott, *Systems Design and the Promotion of Pollution Prevention: Building More Effective Technical Assistance Programs*, 29 GA. L. REV. 449 (1995); Linda Guinn, *Pollution Prevention and Waste Minimization*, NAT. RESOURCES & ENV'T, Fall 1994, at 10; Joan F. Bresnan, *An Analysis of Private, Public and Non-Profit Industrial Pollution Prevention Initiatives*, 16 ENVTL. PROF. 57 (1994).

73. Nor, for that matter, have the states. *See generally* David A. Dana, *The Uncertain Merits of Environmental Enforcement Reform: The Case of Supplemental Environmental Projects*, 1998 WIS. L. REV. 1181, 1182 n.5 (1998) [hereinafter Dana]; Christopher D. Carey, *Negotiating Environmental Penalties: Guidance on the Use of Supplemental Environmental Projects*, 44 A.F. L. REV. 1, 18 (1998) [hereinafter Carey]; Cheryl Hogue, *States, Localities Seek Training to Integrate Pollution Prevention Into Enforcement Work*, Daily Env't Rep. (BNA), Apr. 1, 1997, at D2; *Supplemental Environmental Projects Said Increasingly Used by EPA, States*, Daily Env't Rep. (BNA), June 9, 1994, at D3.

74. Susan Wayland, Assistant Administrator, EPA, Office of Prevention, Pesticides and Toxic Substances, stated at the Spring 1999 Conference of the National Pollution Prevention Roundtable that of the 266 SEPs approved by the Agency in FY 1997, nearly one-half involved pollution prevention obligations. Pollution Prevention: Some Thoughts From the U.S. EPA, National Pollution Prevention Roundtable, Spring Conference, Washington, D.C. (Apr. 1999). On the use of pollution prevention in settlements, *see generally* Jeff Ganguly, *Environmental Remediation Through Supplemental Projects and Creative Negotiation: Renewed Community Involvement in Federal Enforcement*, 26 B.C. ENVTL. AFF. L. REV. 189 (1998) [hereinafter Ganguly]; Lynn L. Bergeson, *EPA Rolls Out a Methodology for Identifying P2 Opportunities for SEPs*, POLLUTION PREVENTION REV., Autumn 1997, at 87 [hereinafter Bergeson]; MONICA M. BECKER & NICHOLAS ASHFORD, U.S. ENVIRONMENTAL PROTECTION AGENCY, RECENT EXPERIENCE IN ENCOURAGING USE OF POLLUTION PREVENTION IN ENFORCEMENT SETTLEMENTS: FINAL REPORT (1995) [hereinafter BECKER & ASHFORD]; U.S. EPA, SUPPLEMENTAL ENVIRONMENTAL PROJECTS UNDER EPCRA SECTION 3131, A REVIEW OF EPCRA SEPS IN REGION 10 (1995); Rachel M. Hopp, *Settling With P2—How Pollution Prevention Initiatives Can Help Your Company Correct Violations and Reduce Penalties*, POLLUTION PREVENTION REV., Autumn 1994, at 387; *Growth Expected in Program to Cut Fines in Exchange for Pollution Prevention*, 23 Env't Rep. (BNA) 2692 (Feb. 12, 1993); *Supplemental Environmental Projects: Pollution Prevention Makes Inroads in Enforcement Settlements*, POLLUTION PREVENTION REV., Winter 1992-1993, at 87. For background on the development of EPA's SEP policy and its application in specific cases, *see generally* David L. Markell, *The Role of Deterrence-Based Enforcement in a "Reinvented" State/Federal Relationship: The Divide Between Theory and Reality*, 24 HARV. ENVTL. L. REV. 1, 14 (2000); Carol Dinkins, *Settlement Issues in Enforcement Actions*, SD88 ALI-ABA 1213 (June 21, 1999); Dana, *supra* note 73, at 1181-82 n.6; Carey, *supra* note 73; Ganguly, *supra*; Cheryle L. Wilson & William L. Thomas, *The ASARCO-EPA Settlement Agreement: Understanding the Value of a Corporate-Wide EMS for Regulators and Strategists Alike*, CORP. ENVTL. STRATEGY, Summer 1998, at 4, 6-9 [hereinafter Wilson & Thomas]; Leslie J. Kaschak, *Supplemental Environmental Projects: Evolution of a Policy*, 2 ENVTL. LAW. 462 (1996); Barnett Lawrence, *Supplemental Environmental Projects: A New Approach for EPA Enforcement*, 26 ELR 10174 (Apr. 1996); Martin Harrell, *Organizational Environmental Crime and the Sentencing Reform Act of 1984: Combining Fines With Restitution, Remedial Orders, Community Service, and Probation to Benefit the Environment While Punishing the Guilty*, 6 VILL. ENVTL. L.J. 243 (1995); Steven A. Herman, *EPA's Revised Supplemental Environmental Projects Policy Will Produce More Environmentally Beneficial Enforcement Settlements*, NAT'L ENVTL. ENFORCEMENT J., July 1995, at 9; Laurie Droughton, *Supplemental Environmental Projects: A Bargain for the Environment*, 12 PACE ENVTL. L. REV. 1141 (1995); Peter Rosenberg et al., *EPA's Revised SEP Policy and the Negotiation of P2 SEPs*, POLLUTION PREVENTION REV., Autumn 1995, at 1.

undertaken voluntarily[75] in settlement of an enforcement action in exchange for a decrease in the amount of the penalty. In this context, the term "environmentally beneficial" means that the project "must improve, protect, or reduce risks to public health, or the environment at large."[76] The 1998 SEP policy requires that there be a nexus, or relationship, between the violation and the proposed project. The project must also advance at least one of the declared objectives of the environmental statute(s) that form the basis of the enforcement action.[77]

The SEP evaluation process is composed of five main steps. EPA will evaluate a proposed project to: (i) ensure that the project meets basic definitional criteria, e.g., that the project is "environmentally beneficial"; (ii) ensure that all legal guidelines, including nexus, are satisfied; (iii) ensure the project fits within an approved category of qualified projects; (iv) determine the appropriate amount of penalty mitigation; and (v) ensure that the project satisfies all implementation and criteria in such areas as liability for performance, oversight and drafting, stipulated penalties, community input, and the like.[78] EPA has identified eight categories of projects that may qualify as SEPs. They are public health; pollution prevention; pollution reduction; environmental restoration and protection; assessments and audits[79]; environmental compliance promotion; emergency planning and preparedness; and an "other" category for worthwhile projects that do not fit within any of the defined categories.[80] A "pollution prevention" SEP, as distinguished from one involving "pollution reduction,"[81] is one that reduces the generation of pollution through source reduction, i.e., any practice which reduces the amount of any hazardous substance, pollutant, or contaminant entering any waste stream or otherwise being released into the environment, prior to recycling, treatment, or disposal.[82] In all cases, the project proponent must demonstrate that there will be an overall decrease in the amount and/or toxicity of pollution, not merely a transfer of pollutant among media.[83] Pollution prevention assessments qualify as SEPs under the category governing assessment and audit projects.[84]

Between FY 1996 and FY 1999, EPA incorporated pollution prevention SEPs in the settlement of 372 enforcement actions. The SEP value of those cases exceeded $303.5 million.[85] Six settlements are summarized below, for purposes of illustration.

- Ashland, Inc. (facility locations: Kentucky, Minnesota, and Ohio; industry sector: oil refining; settlement date: 1998; alleged violations: the Clean Water Act (CWA), RCRA, the Emergency Planning and Community Right-To-Know Act (EPCRA), and TSCA; assessed penalties: $5.8 million; injunctive relief: improve sulfur recovery and reduce particulate emissions, improve wastewater systems to prevent discharges to waters; value of

75. SEPs cannot be used to reward actions required by any federal, state, or local law or regulation. 1998 SEP Policy, *supra* note 63, at B. SEPs can, however, include activities that could be required in the future so long as the regulation does not provide any benefits for early compliance with the prospective requirements. This is a particular concern with air emission reduction programs that may also qualify under the Early Reductions Program under §112(i)(5) of the Clean Air Act (CAA), 42 U.S.C. §7412(i)(5), ELR Stat. CAA §112(i)(5). To the extent that a proposed SEP would convert an early reduction effort into an enforceable commitment in an enforcement settlement agreement, penalty mitigation may be appropriate. *See* Memorandum from Edward E. Reich, Acting Assistant Administrator, EPA, to Regional and Deputy Administrators, Regional Counsels, Air Management Division Directors, and Air and Waste Management Director, regarding Supplemental Environmental Projects in EPA Settlements Involving Early Reductions under the Clean Air Act (Nov. 1, 1991).

76. 1998 SEP Policy, *supra* note 63, at B. While this discussion is limited to applications of EPA's SEP Policy to federal enforcement, many states have similar policies. *See supra* note 73.

77. 1998 SEP Policy, *supra* note 63, at C.

78. *Id.* at A.3; *see id.* at B.-I for further details. With regard to the issue of community involvement in SEPs, see U.S. Environmental Protection Agency, Draft EPA Guidance for Community Involvement in Supplemental Environmental Projects, 65 Fed. Reg. 40639 (June 30, 2000) (seeking comment on community involvement and outlining factors to be considered in evaluating when community involvement in the consideration of a SEP is appropriate).

79. The "assessments and audits" category encompasses pollution prevention assessments, site assessments, environmental quality assessments, and environmental compliance audits. An "environmental compliance promotion" project is one that provides training or technical support to other members of the regulated community to: (i) identify, achieve, and maintain compliance with applicable statutory and regulatory requirements, or (ii) go beyond compliance by reducing the generation, release, or disposal of pollutants beyond legal requirements. Environmental compliance promotion SEPs are acceptable only where the primary impact of the project is focused on the same regulatory program requirements that were violated, and where EPA has reason to believe that compliance in the economic sector would be significantly advanced by the proposed project. *See id.* at D.5 and D.6.

80. The 1998 SEP Policy is, with a few significant exceptions, consistent with the previous 1995 interim policy. One notable difference between the two policy memoranda is that the 1998 SEP Policy places far greater emphasis on the solicitation of public input during SEP negotiations. The 1998 SEP Policy also eliminates EMS audits from the "assessments and audits" category of qualifying projects, making it less likely that SEPs will be based on such projects in the future. If SEP credit is to be given for an EMS under the 1998 SEP Policy, the project will have to be shown to fall within the scope of the "other" project category.

81. The pollution reduction SEP category covers projects that result in a decrease in the amount and/or toxicity of any hazardous substance, pollutant or contaminant entering any waste stream or otherwise being released into the environment by an operating business or facility after the pollutant or waste stream has been generated or released and includes recycling, treatment, containment, or disposal techniques. *Id.* at D.3.

82. *Id.* at D.2. This concept of source reduction is quite broad, and encompasses the following activities: (i) equipment or technology modifications; (ii) process or procedure modifications; (iii) reformulation or redesign of products; (iv) substitution of raw materials; (v) improvements in housekeeping, maintenance, training, inventory control, or other operation and maintenance procedures; (vi) conservation or increased efficiency in the use of energy, water, or other materials; and (vii) "in process recycling," i.e., return of waste products directly to production as raw materials. *Id.* Pollution prevention technologies with particular promise in the SEP context are surveyed in Office of Enforcement and Compliance Assurance, U.S. EPA, Identification of Pollution Prevention (P2) Technologies for Possible Inclusion in Enforcement Agreements Using Supplemental Environmental Projects (SEPs) and Injunctive Relief—Final Report, EPA/300-R-97-001 (1997). *See generally* Bergeson, *supra* note 74, at 87; *Four Pollution Prevention Technologies Could Be Used Widely in SEPs, Report Says*, Daily Env't Rep. (BNA), Apr. 2, 1997, at D9.

83. 1998 SEP Policy, *supra* note 63, at D.2.

84. *Id.* at D.5.A.

85. Some of the 372 cases included SEP components other than pollution prevention. The breakdown by statute is as follows: Clean Air Act (CAA), 52; Federal Water Pollution Control Act, 55; EPCRA, 109; CERCLA, 7; RCRA, 82; Safe Drinking Water Act, 4; TSCA, 42; and Federal Insecticide, Fungicide, and Rodenticide Act, 21. The details of many of these projects can be obtained by searching EPA's SEP database, which can be accessed at <http://es.epa.gov/oeca/sep/searchsep.html>.

SEPs: $14.9 million).[86] The settlement provides for: (i) restoration of 274 acres of rare dune prairie grass in Minnesota and donation of the property to the state; (ii) performance of air monitoring in Kentucky, Ohio, and West Virginia; (iii) installation of hydrofluoric acid release detection and mitigation systems to protect workers and the surrounding communities; (iv) establishment of environmental compliance promotion and education programs for the state of Kentucky; and (v) improvement of oil/water separator and tank systems.

• Koch Industries (facility locations: extensive pipeline system throughout the Midwest; industry sector: petroleum distribution; settlement dates: 1999 and 2000; alleged violations: the CWA and the Oil Pollution Act; assessed penalties: $36 million; injunctive relief: operational and maintenance improvements to prevent future unauthorized releases to the environment; value of SEPs: $7 million).[87] The settlement calls for: (i) a $2 million donation to the Dakota County Park System in Minnesota; (ii) $1.5 million for a conservation project under the Texas Coastal Bend Bays and Estuaries Program; (iii) $500,000 for funding and technical assistance to rural homeowners and schools for alternative wastewater treatment systems; (iv) $350,000 to the Corpus Christi, Texas, Fire Department to purchase an emergency communications center for hazardous materials release response; (v) $150,000 to the Local Emergency Planning Committee to purchase systems for notifying the public of chemical emergencies; (vi) $1 million to conduct a pipeline safety study to educate the oil and gas industry about oil spill prevention; and (vii) $1.5 million for wetlands acquisition and restoration in Kansas and Oklahoma.

• New Jersey Department of Corrections (facility locations: New Jersey; industry sector: correctional facilities; settlement date: 1998; alleged violations: TSCA and polychlorinated biphenyls (PCBs); assessed penalties: $476,196; value of SEPs: $2.2 million).[88] The settlement requires: (i) removal and disposal of PCB transformers; (ii) electrical upgrades; and (iii) environmental assessments of all facilities throughout New Jersey.

• Texaco Chemical Co. (now Huntsman Petrochemicals) (facility locations: Port Neches, Texas; industry sector: industrial organic chemicals facility; settlement date: 1997; alleged violations: Clean Air Act (CAA), CERCLA, RCRA, EPCRA, and TSCA; assessed penalties: $300,000; value of SEPs: $945,000).[89] The settlement calls for the installation of equipment to treat and recover chemicals and wastes previously discharged to the environment.

• Union Oil Company of California (UNOCAL) (facility location: Alaska; industry sector: urea and ammonia plant; settlement date: 1998; alleged violations: CAA; assessed penalties: $550,000; injunctive relief: $6.6 million to reduce ammonia emissions by 1,200 tons per year and substantially reduce risk of catastrophic upsets; value of SEPs: $6.6 million).[90] The settlement requires development and implementation of an operations and maintenance program to monitor combustion efficiency and reduce hydrocarbon and carbon monoxide emissions.

• Weyerhaeuser Company (facility locations: Washington State; industry sector: forest products; settlement date: 1998; alleged violations: EPCRA and CERCLA; assessed penalties: $400,000; injunctive relief: $6 million in operational improvements to avoid repeat violations; value of SEPs: $306,000).[91] The settlement requires: (i) provision of emergency response equipment to local authorities; and (ii) development and operation of an Internet site with current meteorological data to assist in emergency response to future chemical releases.

Because a SEP can be used to mitigate a substantial portion of potential cash penalties, it deserves consideration in the shaping and execution of enforcement settlement strategy.[92] In some cases, however, the costs of SEPs could ultimately prove substantially more expensive than the penalty.[93] Therefore, firms and regulators should undertake a careful cost/benefit analysis before making any commitments, taking into account SEP precedents of a similar type, as well as previous agreements arising under the same statute.

The Use of EMS Principles in Enforcement

The complexity of environmental law and the vigilance of enforcement officials make a plan to attain, and sustain, legal compliance a critical element of management strategy for mid-to-large sized corporations. The dimensions of the challenge vary, depending upon the nature of the business, the industry in which the company competes, and the expectations of its stakeholders. Most firms can, however, reduce the risk of violations and maximize prospects for prosecutorial leniency under the policy and sentencing programs described above through the systematic employment of a variety of precautionary measures, e.g., policies and proce-

86. United States v. Ashland, Inc., No. 98-157 (E.D. Ky., Ashland Division, order entered Jan. 22, 1999) (consent decree).

87. United States v. Koch Indus., No. 1495-118 (S.D. Tex., order entered Jan. 13, 2000) (consent decree).

88. In re New Jersey Dep't of Corrections, No. II-TSCA-PCB-96-0103 (order issued Nov. 12, 1997).

89. In re Texaco Chem. Co., No. MM-92-R6-001-R,T,C,E (order issued Nov. 19, 1998).

90. United States v. Union Oil Co. of Cal. (UNOCAL), No. A97-397CIV(JWS) (D. Alaska, order entered June 8, 1998) (consent decree).

91. In re Weyerhaeuser Co., No. 10-1998-0107 (D. Wash., order issued Mar. 16, 1998) (consent decree).

92. The amount of penalty mitigation is within the Agency's discretion, and will depend upon a consideration of such factors as the actual cash cost of performing the SEP, the benefits of the project for the public and the environment, and any input received from the community. It is never a dollar-for-dollar reduction in penalties. To the extent that a SEP is innovative, the Agency may give more credit for penalty mitigation. The final settlement penalty must equal or exceed whichever is the greater of either (i) the economic benefit of noncompliance plus 10 percent of the gravity component, or (ii) 25 percent of the gravity component only. 1998 SEP Policy, *supra* note 63, at E.

93. *See* Dana, *supra* note 73, at 1200; BECKER & ASHFORD, *supra* note 74, at IV-46.

dures; environmental, compliance, and management system audits; and comprehensive training programs. A significant question for executives is whether, and how, to use an EMS to frame compliance efforts.[94] Recent agreements between industry and the government attest to the importance of a sound EMS for achieving and maintaining compliance with law, and illustrate how companies can use International Organization for Standardization (ISO) 14001, for example, as a component of a larger strategy to prevent or address violations. These agreements also signal a growing awareness within the enforcement community of the positive relationship between such management techniques and conformance with environmental regulations. [Appendix E summarizes several decrees, orders, and settlements requiring features of a compliance-focused EMS.]

In its assessment of the efficacy of extant corporate environmental management approaches, EPA was influenced by its own experiences in recent civil multimedia compliance investigations. A review of a number of recent cases led Steven Sisk and other investigators at EPA's National Enforcement Investigations Center (NEIC) in Denver, Colorado, to conclude that lapses in compliance can often be traced to flawed EMS design or implementation. The NEIC has developed a set of key elements for a compliance-focused EMS model, and this model has informed the design of several settlement agreements, including two that will be described in detail below.

The NEIC articulated the elements of its compliance-focused model, and explained how EMS provisions could be incorporated as components of settlement agreements, in an August 1997, guidance document, revised as recently as January 2000.[95] The NEIC guidance makes it clear at the outset that its EMS model is intended to "supplement, not replace" ISO 14001. In fact, the NEIC compiled the elements of its EMS from a variety of sources, including, but not limited to, ISO 14001 and EPA's 1995 Self-Policing Policy. The basic concept involves using settlement agreements as a vehicle to encourage firms to develop EMS that will not only improve compliance with applicable environmental requirements, but will also enhance environmental performance as measured against goals and objectives set by the organization. The negotiations between ASARCO and the Agency provided the government with an excellent opportunity to translate the guidance into action.

Lessons From ASARCO's P2 and EMS Agreements

Recent settlements between ASARCO and EPA covering operations at various facilities in the United States exemplify the way in which EMS tools and techniques can be utilized.[96] ASARCO and one of its subsidiaries, Encycle/Texas, had pending regulatory disputes and enforcement matters in several jurisdictions. In 1995, ASARCO representatives approached EPA headquarters seeking a forum to resolve all open issues in a comprehensive manner. EPA headquarters and several regions, state agency environmental officials, and ASARCO negotiated commitments that are set forth in three consent decrees, the first two of which were entered into in January 1998 (aggregated for purposes of this Monograph as ASARCO I),[97] and a third, entered into in April 1999 (described for purposes of this discussion as ASARCO II).[98] Read in concert, these settlement agreements trace the outlines of a comprehensive, compliance-focused EMS, bolstered by independent audit and review mechanisms.

ASARCO I

The first agreement between ASARCO and EPA called for the implementation of an EMS affecting its corporate operations throughout the United States. The terms of the agreement are embodied in two consent decrees lodged with U.S. District Courts for the District of Montana and the District of Arizona. These decrees required ASARCO to establish an EMS that encompasses the company's facilities in seven states, and settled alleged violations of more than one environmental statute. EPA and the DOJ consider the agreement a model for other companies in addressing their environmental responsibilities.[99] Arizona Gov. Jane Hull called it one of the most significant environmental settlements in state history.[100] Richard de J. Osborne, chairman of

94. *See* Christopher Bell, *How Does an ISO 14001 EMS Manage Compliance?*, Int'l Envtl. Sys. Update, Mar. 1999, at 15; Carr & Thomas, *supra* note 45; Donald A. Carr & William L. Thomas, *ISO 14001 Sets a Global Environmental Standard*, NAT'L L.J., July 8, 1996, at C2.

95. STEVEN W. SISK, U.S. EPA, OFFICE OF CRIMINAL ENFORCEMENT, FORENSICS AND TRAINING, COMPLIANCE-FOCUSED ENVIRONMENTAL MANAGEMENT SYSTEM—ENFORCEMENT AGREEMENT GUIDANCE, EPA/330-9-97-002 (Aug. 1997). The document was revised in January 2000 to enhance several of the elements and to incorporate more closely the due diligence provisions of the 1995 Self-Policing Policy. *See* Michael J. Penders & Steven W. Sisk, *Trial, Error, and Performance: The Elements of a Compliance-Focused Environmental Management System*, ENVTL. PROTECTION, Mar. 2000, at 20. Under the general EMS framework formulated by EPA investigators for use in settlement agreements, organizations would agree to develop, implement, and maintain a management system addressing the following key elements: environmental policy; organization, personnel, and oversight of EMS; accountability and responsibility; environmental requirements; public involvement/community outreach; maintenance of records and documentation; assessment, prevention, and control; environmental incident and noncompliance investigations; environmental training, awareness, and competence; environmental planning and organizational decisionmaking; pollution prevention program; and continuing program evaluation and improvement. For a deeper analysis of the EPA compliance-based EMS and similarities and differences between it and ISO 14001, *see generally* Cheryle L. Wilson, *Preventing and Mitigating Environmental Enforcement With an EMS*, POLLUTION PREVENTION REV., Summer 1998, at 29.

96. The discussion of the ASARCO-EPA environmental management system agreements builds on and updates a previous, and in some respects more detailed analysis, presented in Wilson & Thomas, *supra* note 74.

97. *See* United States v. ASARCO Inc. (D. Mont. May 5, 1998 and D. Ariz. June 2, 1998) (consent decrees).

98. *See* United States v. Encycle/Texas, Inc., and ASARCO Inc. (S.D. Tex. Oct. 6, 1999) (consent decree).

99. *See, e.g.*, Steven A. Herman, *Innovations in Environmental Enforcement and Compliance*, NAT'L ENVTL. ENFORCEMENT J., February 1999, at 3, 8; Lois J. Schiffer & Timothy J. Dowling, *Government Works! Case Studies in Environmental Protection*, 32 CREIGHTON L. REV. 761, 790 (1999) ("The [ASARCO] settlement demonstrates that even after lawsuits are filed, cooperation between government and industry can resolve the litigation and address the continuing needs of the environment."); Steven A. Herman, *Innovative GTE, ASARCO Settlements Enhance Prospects for Future Compliance While Correcting Past Violations*, NAT'L ENVTL. ENFORCEMENT J., Mar. 1998, at 12.

100. Jane Larsen, *ASARCO Cleanup Bill: $62 Million; Copper Giant to Fix Problems in 2 States*, ARIZ. REPUBLIC, Jan. 24, 1998, at A1. For further discussion of the agreement, *see ASARCO to Spend $61.5 Mil on Environmental Cleanup*, Metals Wk., Jan. 26, 1998, at 3; Bob Ortega, *ASARCO Will Pay $6.4 Million Fine in Pollution Case*, WALL ST. J., Jan. 26, 1998, at B6.

ASARCO's Board of Directors and the company's (now retired) chief executive officer, was also positive, if more restrained, in his assessment that the agreement is "a basis for the Company and Federal and State agencies to work together in the future."[101]

The Montana decree [excerpts of which are attached] resolved disputes over alleged violations of RCRA and the CWA by the company's East Helena, Montana, lead smelter, while the Arizona decree addressed injunctive relief and civil penalty claims of the federal government under the CWA for alleged permit violations and unpermitted discharges at the company's Ray copper mine complex.[102] Choosing to negotiate rather than litigate these issues, the parties crafted a settlement that provides for—in addition to a new companywide EMS—implementation of compliance measures, corrective action, and several SEPs[103] (including wetlands and wildlife habitat improvements at the lead smelting facility in Montana), environmental protection measures at ASARCO's operating facilities, implementation of an extensive compliance plan at the company's Ray copper mine, and payment of substantial civil penalties. By the time it has fulfilled its obligations under the terms of its agreement with EPA, ASARCO will have paid $6.38 million in penalties and expended over $50 million to reduce suspected impacts of heavy metals, including arsenic, mercury, and lead, to surface and groundwater.

To honor its commitment to conduct all operations and activities "responsibly and in a manner designed to protect the health and safety of its employees, its customers, the public and the environment," ASARCO designed a multifaceted EMS modeled upon ISO 14001.[104] The ASARCO EMS incorporates state-of-the-art computer technology, internal compliance and management system environmental audits, employee environmental awareness training, and close communications with communities and governmental agencies. The elements of this system are described in the ASARCO Management System Standard, or AMS Standard, which contains corporatewide management system standards for environmental and other key business objectives. The significance of this system is emphasized in the company's EMS policy, which also contains supplemental requirements for the environmental component of the AMS Standard. The EMS policy is applicable to all mines, plants, operating or service groups or departments within the company[105] that make products or provide services and that have or can have a significant environmental impact, and requires them to: (i) implement, maintain, and improve an EMS; (ii) assure conformance with ASARCO's stated environmental policy; and (iii) demonstrate such conformance to others.

The ASARCO EMS is structured around five main elements: (i) environmental policy; (ii) planning; (iii) implementation and operation; (iv) checking and corrective action; and (v) management review. Under ISO 14001, an organization's environmental policy must include a commitment to compliance with applicable laws. The EMS policy charges ASARCO's management committee with the responsibility of defining an environmental policy and with ensuring that it includes, among other things, a commitment to continual improvement and prevention of pollution and a commitment to comply with relevant environmental legislation and regulations and with other requirements to which ASARCO subscribes. This environmental policy is to be documented, implemented, maintained, and communicated to all employees, and must also be made available to the public. ASARCO's Environmental, Safety and Health policy fulfills these obligations. It also sets forth the company's commitment "not only to comply with existing environmental, safety and health laws and regulations but also to support additional activities that contribute to environmental protection, responsible resource management and the safety and health of employees, customers and members of the community."

ASARCO's EMS planning process calls for the establishment of procedures to identify environmental aspects of the company's activities, products, or services, in order to determine those that have or can have significant environmental impact. A determination concerning the significant environmental aspects of a particular unit must be made where the unit has yet to do so, prior to the start-up of a new process or product, or prior to modification of an existing process or product which creates new environmental aspects or significantly increases existing environmental impacts. In addition, under the EMS policy company units are required to create and follow procedures to identify and have access to legal and other requirements directly applicable to environmental aspects of their activities, products, or services. The system further mandates that a "comprehensive compliance obligations calendar" be prepared and kept updated.

101. ASARCO, ASARCO Enters Compliance Initiative Agreement With EPA (Jan. 23, 1998) (press release) (quoting Richard de J. Osborne).

102. The Ray copper mine in Hayden, Arizona, and the lead smelting plant in East Helena, Montana, play important roles in the ASARCO empire. The Ray mine is part of a larger operation, consisting of a 265,000 ton/day open pit mine, a 12,000 ton/day silicate heap leaching system, a low grade sulfide dump leaching system, a solvent extraction-electrowinning operation, and associated maintenance, warehouse and administrative facilities. The East Helena smelting operation, on which the company has expended more than $100 million on environmental improvements over the last 10 years, is also significant, processing 300,000 tons of material to produce 70,000 tons of lead bullion annually. Jim Ludwick, *Older Factories Can Still Turn a Profit*, BILLINGS GAZETTE, Sept. 22, 1997, at A1. The facility treats ores and other lead bearing materials from small mines in Montana and from as far away as Asia.

103. The SEP elements in the agreement between ASARCO and EPA conform to the Agency's Interim Revised EPA Supplemental Environmental Projects Policy. 60 Fed. Reg. 24856 (May 10, 1995), ADMIN. MAT. (available from the ELR Document Service, ELR Order No. AD-3283). *See generally* Steven A. Herman, *EPA's Revised Supplemental Environmental Projects Policy Will Produce More Environmentally Beneficial Enforcement Settlements*, NAT'L ENVTL. ENFORCEMENT J., July 1995, at 9. The interim policy was later superceded by the 1998 SEP Policy, discussed *supra* note 63.

104. For further information on the ISO's specification on EMS, ISO 14001, and other standards in the ISO series, *see generally* ALAN SCHOFFMAN & ALAN TORDINI, ISO 14000: A PRACTICAL APPROACH (Oxford Univ. Press 2000); THE ISO 14000 HANDBOOK (Joseph Cascio ed., American Soc'y for Quality 1999); GAYLE WOODSIDE ET AL., ISO 14001 IMPLEMENTATION MANUAL (McGraw Hill 1998); S. WAYNE ROSENBAUM, ISO AND THE LAW (AQA Co. 1998); IMPLEMENTING ISO 14000 (Tom Tibor & Ira Feldman eds., Irwin 1997); Christopher L. Bell, *The ISO 14001 Environmental Management Systems Standard: A Modest Perspective*, 27 ELR 10622 (Dec. 1997); Carr & Thomas, *supra* note 45, n.160 (containing bibliography of materials published as of 1997); Christopher L. Bell, *ISO 14001: Application of International Environmental Management Systems Standards in the United States*, 25 ELR 10678 (Dec. 1995). For a status report on the spread of ISO 14001 globally, see Charles Corbett & David Kirsch, International Diffusion of ISO 14001 Certification (Jan. 2000) (unpublished paper, available at <http://www.anderson.ucla.edu/Acad_unit/otm/research.htm>).

105. The agreement covers all ASARCO operations in the United States except for certain subsidiaries and joint ventures.

The ASARCO system adopts the ISO 14001 rubric of environmental objectives and targets, requiring units to establish, maintain, and document such objectives and targets at each relevant function and level within the unit. Some parties within and outside EPA have previously expressed the concern that if ISO 14001 were taken literally, an organization could establish a procedure without implementing it and still obtain certification. ISO 14001 requires only that an organization "consider" significant aspects, legal requirements, and the views of interested parties when setting its objectives and targets. Under ASARCO's EMS, units of the company must develop a documented environmental action plan designating responsibility for the achievement of objectives and targets, and the means and time frame by which they are to be achieved. Similarly, whereas ISO 14001 §4.4.3 simply requires that an organization "consider processes for external communication on its significant environmental aspects and record its decision,"[106] the ASARCO EMS requires that procedures for such communications be established and maintained, that they be established at the unit level, and that the communication include information on ASARCO's regulatory compliance status. Furthermore, unlike ISO 14001, the ASARCO EMS mandates establishment and maintenance of procedures for external communication with suppliers, on-site service providers, and contractors.

Management of each unit of ASARCO is responsible for developing, implementing, and maintaining the EMS, including the appointment of a management representative. The management representative ensures that the unit's EMS is properly established, implemented, and maintained, and to report on the unit's compliance with the EMS to unit management. Training is an important facet of implementation and operation of the system. Personnel whose work may create a significant environmental impact must receive appropriate training, with specific attention being given to subjects such as applicable legal requirements and the potential consequences of departure from specified operating procedures. As part of system implementation and operation, units must also establish and maintain procedures covering internal and external communications.

The ASARCO management system incorporates a number of mechanisms to measure, monitor, and evaluate environmental performance at the unit level.[107] The management system provides for periodic, internal audits of the EMS and goes beyond ISO 14001 in explicitly requiring audits of compliance with applicable environmental laws and regulations. Actual or potential nonconformity with either the EMS or legal requirements must be investigated and corrected in accordance with the requirements of corporate ASARCO standards. The EMS itself is to be reviewed by ASARCO management periodically to ensure continuous improvement. The management review includes an evaluation of the unit's performance as measured against its objectives. Records of such reviews are to be maintained for an indefinite period.

ASARCO II

Another consent decree with EPA and the Texas Natural Resources Conservation Commission (TNRCC), approved by the U.S. District Court in Houston, Texas, in late 1999 broadens "the ASARCO-initiated [EMS] to establish Company-wide environmental audit procedures,"[108] and requires ASARCO to pay $5.5 million, as well as undertake a set of SEPs.[109] Under ASARCO II [excerpts of which are attached], the company agreed to develop and implement a five-year EMS audit program to assess whether an effective EMS is being implemented at operations covered by the decree, commencing in January of the year 2000. Phase I activities (audit years one and two, e.g., 2000 and 2001) are intended to provide the company with feedback on EMS implementation progress. During this period, audits will encompass: (i) verification of progress on elements addressed by each site's EMS Development Plan; (ii) evaluation of, where systems have been implemented, conformance to the AMS Standard and the ASARCO EMS policy; and (iii) a determination of whether the company is effectively communicating environmental requirements to affected parts of the organization, including contractors.

In Phases II and III of the ASARCO EMS audit program (years three, four, and five, or 2002, 2003, and 2004), the company will: (i) assess the EMS at each site to determine whether its implementation meets the criteria defined by the AMS Standard and the ASARCO EMS policy; (ii) identify any issues that may require further improvements or evaluations; and (iii) determine if the EMS is being maintained, and that "continuous improvement has been institutionalized." These audits are to be conducted in accordance with ISO 14011. In analyzing the adequacy of the company's EMS implementation effort, the auditors will focus on whether: (i) there is a defined program or planned task for the EMS element; (ii) the program or task has been implemented, is being maintained, and to what extent; (iii) each site's internal self-assessment procedure for programs and tasks composing the EMS are adequate; (iv) the company is

106. Compare with ISO 14001 §4.4.6, requiring an organization to establish and maintain procedures related to the identifiable significant environmental aspects of goods and services used by the organization and the communication of relevant procedures to suppliers and contractors.

107. ISO 14001 requires monitoring and measurement of the key characteristics of the organization's operations and activities that can have a significant impact on the environment. Moreover, organizations are expected to establish and maintain a documented procedure for periodically evaluating compliance with relevant environmental legislation and regulations. On the relationship between ISO 14001 and compliance strategy, see Bell, *supra* note 94, at 15; Carr & Thomas, *supra* note 45. ISO 14004 states that a system for measuring and monitoring actual performance against the organization's environmental objectives and targets includes an evaluation of compliance with relevant environmental legislation and regulations. The process for performing the requisite periodic evaluation of compliance, however, is left unspecified. Also unexplained is what is meant by the term "periodic," a term that is also left undefined in EPA's 1995 Self-Policing Policy.

108. ASARCO, ASARCO Concludes Compliance Initiative Agreement With U.S. EPA (Apr. 15, 1999) (press release).

109. The agreement requires incorporation of recycling into the company's programs. In El Paso, when the smelter reopens ASARCO will replace a portion of the coke used in the furnace with scrap tires (1,200 tons per year). The Corpus Christi project will expand technological innovations that improve the electrowinning process to recover metals from waste materials at the commercial recycling operation of Encycle/Texas. The SEP in Corpus Christi is a conservation easement to maintain land owned by the company in the community in its natural state in perpetuity. The El Paso SEP is designed to reduce air particulate matter through ASARCO's expenditure of $1.850 million to pave roads, alleys, parking lots, and medians. The Tennessee project, called the "Coy Mine SEP," entails the restoration of riparian and wetland values.

effectively communicating environmental requirements to affected parts of the organization; (v) further improvements should be made to the EMS; and (vi) there are observed deviations from ASARCO's written requirements or procedures. Auditors performing the EMS audits and drafting audit reports must satisfy requirements of ISO 14012. Classroom training must include successful completion of an ISO 14000 auditing course.

By the end of Phase III, the company must conduct an EMS audit of each covered operation, contracting with a qualified, independent, third-party auditor to conduct these audits. Auditor must satisfy qualification requirements of ISO 14012. In addition to EMS issues identified above, the contract auditor may also report on areas of concern that, in their judgment, merit further review or evaluation for potential environmental or regulatory impacts.[110] In the case of EMS audits in Phases II and III, the company must prepare a written EMS audit plan, audit report, and EMS audit action plan for each covered operation. ASARCO must also, for Phase II and III audits, prepare an annual report listing the dates and locations of audits, personnel conducting the audit, and the date the action plan was drafted.[111] The ASARCO II agreement called for a similarly comprehensive compliance environmental audit program that includes third-party verification of the company's conformance with the agreement.[112]

EPA's Model EMS

EPA's stated purpose in requiring ASARCO to fulfill these obligations is to enhance the company's development of a comprehensive corporate-level EMS to promote compliance with all environmental requirements, achieve pollution prevention, and accomplish pollution reduction. To appreciate what Administrator Carol Browner meant by use of the term "model" in the context of the ASARCO settlement, it is useful to examine the way in which the consent decrees reinforce the ASARCO EMS.

The environmental management and pollution requirements under the ASARCO consent decrees are divided into two categories, "environmental compliance promotion" and "pollution prevention and pollution reduction." The environmental compliance promotion category encompasses employee environmental awareness training, development of a protocol for identifying and communicating environmental requirements, and the sharing of the protocol, along with any "lessons learned," with other businesses and organizations in a series of three seminars to occur in each of the fourth and fifth years after entry of the decrees by the district court. These "technology transfer" sessions, which ASARCO is obligated to arrange, promote, and lead (with or without outside assistance), are to focus on such issues as how potentially applicable current and proposed environmental requirements are systematically identified, and how applicable requirements are integrated into ASARCO facility activities and processes.

The pollution prevention and pollution reduction provisions of the agreement obligate ASARCO to develop an EMS implementation plan containing implementation milestones for each ASARCO facility. ASARCO must also undertake an environmental aspects/environmental impacts review. Based on this review, ASARCO is to implement an assessment, prevention, and control program including, among other things, processes for the following activities: (i) ongoing assessments of operations for such purposes as preventing and controlling releases, environmental protection, and environmental compliance (including monitoring and measurements to maintain compliance and root cause analysis of problems); (ii) identification of activities that could cause adverse environmental impacts and/or regulatory noncompliance, and where standard operating procedures need to be developed; (iii) routine, internal self-inspections; and (iv) development and implementation of corrective action procedures which address specified internal/external indicators of actual and/or potential noncompliance (e.g., a notice from a government agency of alleged noncompliance or an observed or documented instance of noncompliance with EMS policies or procedures). In addition, ASARCO is required to collect data and report annually on a specified set of environmental metrics, including spills and accidental releases, permit exceedances, toxic and pollutant releases, hazardous waste generation, recycling, and water and energy usage to measure the efficacy of its programs.

The ASARCO settlements illustrate one firm's approach to legal compliance in the context of an ISO 14001-type EMS. ASARCO had developed a blueprint for overhauling its corporate management, which included its EMS (and several meaningful components relating to compliance). These steps were taken before it entered into negotiations with EPA over enforcement issues at its facilities. This made it easier for the parties to come to terms on a package of environmental management and pollution prevention requirements. Measures in ASARCO's EMS that some firms might find worthy of emulation include the comprehensive compliance obligations calendar, the inclusion of legal and other requirements as part of the impact/aspect analysis, and procedures to keep employees, on-site service providers and subcontractors abreast of relevant legal and other requirements. Also, the regularly scheduled self-inspections of both the EMS and underlying compliance program elements may contribute not only to stronger overall environmental management, but also to more consistent compliance with applicable regulations.

ASARCO's EMS requires completion of an analysis concerning its activities, products, and services that have an impact on the environment. As described above, ASARCO I prescribed the methods for undertaking such an analysis in exacting detail. The major components of this analysis comprise: process flow diagrams and associated descriptive information for each ASARCO (and on-site contractor) operation at every ASARCO operating facility, ongoing assessments of operations to include self-inspections and root cause analysis of identified problems, and corrective action

110. The company must provide EPA with a schedule of the EMS audits to be performed during Phases II and III.

111. This document must also contain a certification by a responsible corporate officer that all required audits were performed in accordance with the decree.

112. Examples of the imposition of rigorous compliance management obligations may be found in the comprehensive compliance program agreed to by American Airlines in connection with its plea agreement in the U.S. District Court of the Southern District of Florida in December 1999 (on file with authors), and the Environmental Compliance Plan prepared pursuant to the plea agreement between Royal Caribbean Cruises, Ltd. and the United States, filed in the U.S. District Courts for the District of Puerto Rico and the Southern District of Florida in June 1998 (on file with authors).

procedures. Other firms can draw insights from this approach, as well as from a detailed, process-based investigation guide available from NEIC.[113]

The ISO 14001 requirement that an organization simply "consider" processes for external communication concerning significant environmental aspects has always been controversial and has struck some environmentalists and some state regulators as inadequate.[114] In certain situations, firms may find community involvement and outreach to be an effective method for encouraging adherence to best environmental management practices and accomplishing other important objectives.[115] A program for external communication like that of ASARCO's, that educates the community on the environmental impacts of its activities, products and services, could serve as a useful prototype for other firms.

Training, whether of employees, contractors, or other on-site personnel, is a central component of ASARCO's EMS. The company's first settlement agreement with EPA also emphasizes this area. The ASARCO EMS and the consent decrees concentrate on three areas of training in particular. First, there must be annual awareness training for employees at all levels regarding their roles and responsibilities under company policy, procedures, and EMS requirements. In addition, workers must be made cognizant of the actual or potential significant environmental impacts of their activities. Finally, affected employees, subcontractors, and on-site service providers must be educated about the legal and other requirements applicable to their respective jobs. Generally, a company training program may satisfy ISO 14001 and EPA compliance-focused EMS requirements if the program provides for systematic, companywide application, regular review, and incorporation of results into program improvements.

A third-party audit is only required when a firm chooses to be certified to the ISO 14001 standard, which requires an audit by an accredited registrar. The specification does, however, require periodic, internal EMS audits. EPA has attempted to shore up what it perceives to be the limitations of ISO 14001 through the imposition of supplemental comprehensive external auditing obligations. One can expect enforcement officials will continue to pursue the inclusion of comparable provisions in future agreements, and the agreements negotiated to date at the federal and state level illustrate varying approaches to such a tactic.

An Emerging Trend

The ASARCO outcome reflects a modest trend in the creative use of consent decrees, pleas, and other enforcement agreements to promote better environmental compliance through management systems implementation and improvement. Indeed, the number of such agreements is growing as states begin to experiment with EMS requirements in their own enforcement programs.[116] In many of the cases at both the federal and state level, ISO 14001 serves as a key point of reference.

As recently as September 1999, BP Exploration (Alaska) Inc. (BPXA or the Company) entered into a plea agreement with the U.S. Attorney's Office for the District of Alaska and the DOJ, pursuant to which the Company will implement an EMS throughout its U.S. oil exploration, drilling, and production operations.[117] The case arose out of an investigation of waste disposal practices of BPXA's contractor, Doyon Drilling, at the Company's Endicott oil field on Alaska's North Slope between 1993 and 1995.[118] BPXA learned of the alleged illegal disposal on August 31, 1995, when a Doyon employee brought it to the attention of BPXA's Shared Service Drilling organization. BPXA reported the violations two weeks later, following an internal investigation that found the whistleblower's allegations to be credible. That two-week delay, though brief, would prove costly.

After a protracted investigation involving a sequence of grand juries, BPXA pled guilty to one felony count of knowingly failing to immediately report a release into the envi-

113. STEVEN W. SISK ET AL., U.S. EPA, OFFICE OF ENFORCEMENT AND COMPLIANCE ASSURANCE, NATIONAL ENFORCEMENT INVESTIGATIONS CENTER, PROCESS-BASED INVESTIGATION GUIDE, EPA/330-9-97-001 (1997).

114. An assessment by the Technology, Business and Environmental Group at MIT found that firms are not likely to involve outside stakeholders in the ISO goal-setting process unless they were already doing so prior to registration. Jason Switzer et al., *ISO 14001 and Environmental Goal Setting: Promises Kept*, ENVTL. QUALITY MGMT., Winter 1999, at 1, 19. Many regulators and environmental organizations are calling for revision of ISO's EMS specification to clarify the area of external communications. *See generally Once and for All, ISO 14001 Will Be Revised: But Experts Agree, No Additional Requirements in New Standard*, Int'l Envtl. Sys. Update, July 2000, at 1. For a view of how clarification of the standard might affect its implementation, see Christopher Bell, *Clarifying ISO 14001—The Road to a Better Standard?*, Int'l Envtl. Sys. Update, June 2000, at 14.

115. *See generally* CAROL J. FORREST & RENEE HIX MAYS, THE PRACTICAL GUIDE TO ENVIRONMENTAL COMMUNITY RELATIONS (John Wiley & Sons 1997); William Looney, *What the Shell Brent Spar Incident Means for Strategists and Other Corporate Planners*, CORP. ENVTL. STRATEGY, Autumn 1997, at 5.

116. Authorities in Canada have begun to employ similar measures. For example, a provincial court in Alberta ordered Prospec Chemicals Ltd. to obtain ISO 14001 certification after the company was charged with exceeding allowed air emissions of total reduced sulfur. Initially, Prospec had proposed ISO 9001 implementation, and prosecutors countered with ISO 14001. Prospec ultimately decided to set up an integrated management system incorporating ISO 9002, ISO 14001, and BS8800, an occupational health and safety management standard. *See generally* Robert Abbott et al., *Making ISO Work at Mid-Size Firms: The Prospec Chemicals Ltd. Case*, CORP. ENVTL. STRATEGY, Autumn 1997, at 73. Another firm, Coretec Inc., worked out an arrangement with the Ontario Court (provincial division) Toronto Region in August 1998 under which Coretec would address alleged violations of the Canadian Environmental Protection Act by meeting specified EMS requirements, i.e., establishing an environmental policy; identifying the environmental aspects arising from the company's past, existing or planned activities, products or services to determine all environmental impacts; identifying and complying fully with the relevant legislative and regulatory requirements; identifying environmental priorities and setting appropriate environmental objectives and targets; establishing programs to implement the policy and achieve objectives and targets; facilitating planning, control, monitoring, corrective action, auditing and review activities to ensure that the policy is complied with and that the EMS remains effective; and finally, obtaining ISO 14001 certification. *See generally Environmental Management System for Company Ordered by Canadian Court*, Daily Env't Rep. (BNA), Aug. 27, 1998, at A-7.

117. United States v. BP Exploration (Alaska) Inc. (D. Alaska Sept. 1999) (plea agreement). *See generally BP Exploration Averts RCRA Charges With Environmental Management System*, Daily Env't Rep. (BNA), Feb. 16, 2000, at A-3; Oil Giant Fined $22 Million, Int'l Envtl. Sys. Update, Nov. 1999, at 3; Don Hunter, *Dumping to Cost BP $22 Million*, ANCHORAGE DAILY NEWS, Sept. 24, 1999, at 1.

118. The contractor had intermittently disposed of used oils, solvents, glycols and possibly paint thinners by injecting them into the outer annuli of various Endicott wells together with drilling muds and other drilling liquids. Interview with Chris Phillips, Business Unit Leader—Shared Services Technology, BP Exploration (Alaska) Inc. (Dec. 21, 1999) [hereinafter Phillips Interview].

ronment of a hazardous substance in violation of CERCLA, paying a criminal fine of $500,000. Federal prosecutors agreed to forgo charging the company with criminal violations of RCRA, and to not seek additional prosecution for related allegations of environmental criminal violations and general criminal violations, in light of BPXA's adoption of corrective measures subsequent to the offense, its taking certain steps to cooperate with the government's investigation, and its development, implementation, and maintenance of a nationwide EMS. The Company also agreed to a civil settlement under which it would pay a penalty of $6.5 million to resolve allegations that the Company had illegally disposed of hazardous waste and also violated the Safe Drinking Water Act.[119] The Company's Endicott Island operation had received ISO 14001 certification in December 1998.

Under the plea agreement and a document attached to the plea agreement entitled *BP Amoco Group Environmental Management System*, BPXA committed to "develop, implement and maintain" a nationwide EMS at BP Amoco Group facilities engaged in the exploration, drilling, and/or production of oil, both onshore and offshore, in the United States and its territories.[120] Such a commitment struck the Company as appropriate given BPXA's conclusion that the "violations occurred because our systems for ensuring environmental compliance weren't working as well as we believed."[121] The Environmental Law Institute and the University of North Carolina at Chapel Hill are engaged in a multi-year project involving the creation of the National Database on Environmental Management Systems (NDEMS). NDEMS includes more than 50 pilot facilities that are implementing EMSs. The joint ELI-UNC project is supported by the EPA and a consortium of states known as the Multi-State Working Group. The project will examine the effect EMSs have on environmental performance, economic performance, compliance with regulatory requirements, pollution prevention, and involvement with external interested parties. NDEMS data collection and research will be continuing for several more years in order to analyze the effects of EMS implementation. In at least one instance, parties have negotiated voluntary provision of data to this database as a component of an enforcement settlement.[122] The EMS must be "consistent with the best environmental practices in order to effectively protect workers, the public and the environment, and to comply with all statutory and regulatory requirements."[123] The design of the EMS is primarily derived from the NEIC's EMS guidance, but with additional input from internal and external sources, such as BP Amoco principles and policies, ISO 14001,[124] the DOJ's *Factors* Policy, and EPA's 1995 Self-Policing Policy. The system embodies an "enhanced emphasis on compliance assurance, pollution prevention and adequate contractor and on-site service provider supervision."[125] Some of the specific elements to be covered in the BP Amoco Group EMS are noted below.

- Management Policies and Procedures. Among other points, the environmental policy must clearly communicate BP Amoco Group's "commitment to pollution prevention and continual improvement" of the EMS.
- Organization, Personnel, and Oversight. The EMS must, among other things, include organization charts that identify units, line management, and other individuals having responsibility for compliance with environmental requirements.
- Accountability and Responsibility. The EMS must, among other things: (i) specify accountabilities and responsibilities of management to adequately supervise the compliance with environmental, safety and health requirements by contractors and on-site service providers; (ii) measure employee and manager environmental performance (in quantitative terms); and (iii) reward positive environmental performance of employees, and con-

119. For its part, Doyon Drilling entered a plea agreement with the DOJ in April 1998. The contractor pled guilty to 15 violations of the CWA and accepted deferred prosecution under 2 other environmental statutes. Three of its employees also pled guilty, and one received a one-year prison sentence. Phillips Interview, *supra* note 118.

120. The EMS is to be implemented and maintained at selected BP Amoco Group Business Units owned, managed, or operated by BPXA, BP Exploration & Oil, and/or Amoco Production Company. The plea agreement also requires BPXA to make modifications to its employee "Gainshare" (performance bonus) program, and to its contractor incentive bonus scheme, at covered operations.

121. BP, BP Exploration (Alaska) Reaches Endicott Island Plea Agreement With U.S. Justice Department (Sept. 23, 1999) (press release) (quoting Richard Campbell, president of BP Exploration (Alaska) Inc.).

122. On June 5, 2000, ELI and UNC released the current NDEMS database as well as a landmark study that contains a thorough analysis of the initial baseline data from the 50 facilities. The report, The Effects of Environmental Management Systems on the Environmental and Economic Performance of Facilities, covers the three years prior to the implementation of an EMS at each facility, including extensive data on each facility's past environmental performance, compliance history, pollution prevention efforts, and involvement of external interested parties. The facilities, including some that are implementing EMSs based on the ISO 14001 standard, are located in 10 states and represent a variety of sizes, industries, and government agencies.

In addition to analyzing the baseline data, The Effects of Environmental Management Systems includes a summary of the major project mileposts and updated information on the demographics for the 50 pilot facilities. The report also describes upcoming project highlights, including preliminary details regarding the facilities' EMS design activities. Beyond analyzing information from the national database, the ELI-UNC researchers have conducted a number of site visits and in-person interviews with pilot facility managers and employees. Two facilities are the subject of detailed case studies in the report, and more case studies will be published during 2000. In the next phase of the NDEMS project, the ELI-UNC research team will be compiling data on the pilot facilities' EMS designs, as well as the processes by which their EMSs are implemented. These data will be added to NDEMS, which will be updated every six months so that changes over time can be analyzed by the ELI-UNC team and other researchers. The data base and the report are available at www.eli.org/isopilots.htm. The project website also includes background information on NDEMS, including the data collection protocols.

123. BPXA further agreed to assume all costs for the development, implementation, and maintenance of the EMS during the five-year probation period, estimated at approximately $15 million. The company is thought to have spent approximately $5 million to implement the present EMS on Endicott Island and in Alaska.

124. Under the agreement with federal prosecutors, each BP Amoco Group Business Unit EMS must be consistent with the following ISO 14001 principles: (i) line management and responsibility; (ii) clear roles and responsibilities; (iii) competence commensurate with responsibilities; (iv) balanced priorities; (v) identification of environmental hazards; (vi) operational and hazard control; (vii) checking and corrective action; and (viii) continual improvement. The agreement does not mandate ISO 14001 certification, though ultimately many of the facilities covered under the agreement could become certified consistent with the environmental goals of the company.

125. *Id.*

tractor and on-site service providers through incentive bonus programs and shared savings programs (based on input metrics, e.g., year-on-year improvement in overall air or water emissions per unit of production).

• Environmental Requirements. The EMS must, among other things, describe a process for identifying, interpreting and effectively communicating environmental requirements to affected employees, on-site service providers, and contractors.

• Assessment, Prevention, and Control. The EMS must, among other things: (i) identify an ongoing process for assessing operations, for the purposes of preventing and controlling releases, ensuring environmental protection, and maintaining compliance; the process must include performing root-cause analysis of identified problems in order to develop remedies and prevent recurring issues; (ii) describe a system for conducting and documenting routine, objective self-inspections; (iii) describe a system for conducting systematic, objective, periodic review of facility operations and practices related to meeting environmental requirements by qualified, independent auditors; (iv) describe and demonstrate top management commitment to compliance; and (v) describe a compliance auditing program which includes quality assurance component whereby an independent objective, qualified third-party organization audits the compliance audit program, reporting directly to top management at least once every two years.

• Environmental Incident and Noncompliance Investigations. The EMS must, among other things, implement a process for contractor and on-site service provider employees to notify management without fear of reprisal regarding environmental compliance issues and concerns.

• Environmental Training, Awareness, and Competence. The EMS must, among other things, describe a process for tracking employee, contractor, and on-site service provider questions about interpretation of environmental regulations.

• Planning. The EMS must, among other things, describe how environmental planning will be integrated into other plans developed by the organization.

• Maintenance of Records and Documentation. The EMS must, among other things, identify the types of records developed in support of the EMS, including self-audits, third-party audits and reports, audit working papers, and correspondence, including emails, about audits.

• Pollution Prevention. The EMS must describe an internal program for "preventing, reducing, recycling, reusing and minimizing waste and emissions."[126]

• Continuing Program Evaluation and Improvement. The EMS must, among other things: (i) describe a program for periodic (at least annual) evaluation of the EMS; and (ii) describe a program for periodic (at least every two years) third-party evaluation of the EMS.

• Public Involvement/Community Outreach. The EMS must describe a program for ongoing community education and involvement in the environmental aspects of the organization's products, activities, and services.

The plea agreement also calls for BP Amoco Group to establish certain monitoring and oversight functions.[127] BP Amoco Group will appoint three compliance committees (Corporate Compliance Committee), composed of environmental, engineering, drilling and operating personnel, corporate legal and environmental staff, and an independent third-party consultant with expertise in designing EMS. One committee will oversee business units involved in the exploration, drilling, and production of oil onshore and offshore in Alaska[128]; a second committee will oversee business units involved in such activities in the contiguous United States; and a third will oversee business units involved in such activities located offshore the contiguous United States.

Each Corporate Compliance Committee will evaluate the current EMS within each business unit in order to (i) compare the current EMS to the EMS objectives and elements set forth in the plea agreement and determine what, if anything, is needed to meet these objectives and key elements; (ii) identify strengths and weaknesses of the business unit's current EMS; and (iii) evaluate the effectiveness of the current EMS to address present and anticipated environmental issues and risks. These evaluations will inform each committee's preparation of an EMS for each business unit, complemented by an EMS manual. The Alaska Corporate Compliance Committee must submit draft and final versions of the Endicott EMS manual to EPA Region 10 for the Agency's review, comment, and approval.[129] In the case of the Endicott EMS, the Alaska Corporate Compliance Committee must also arrange and fund an EMS compliance audit[130] by an independent third party, with both the scope of such audit and the independent auditor to be pre-approved by EPA Region 10. The initial report of the auditors is to be submitted simultaneously to EPA Region 10 and BP Amoco Group. In the case of EMS other than that implemented at Endicott, the Corporate Compliance Committees must re-

126. The company anticipates developing clearer guidance for field personnel on how to manage waste streams with the expectation such measures will lead to improved disposal conformance and reduced pollution. Phillips Interview, *supra* note 118.

127. The probation officer is responsible for ensuring that BPXA conforms to the plea agreement. However, the agreement states that a court-appointed monitor will "monitor the development, implementation and maintenance of the EMS and the design and supervision of the incentive bonus and shared savings programs" on the behalf of the probation officer. BPXA agreed to provide the government with the names of three candidates qualified, by virtue of experience and expertise, for the position of monitor; the DOJ will select one of these candidates. BPXA will be responsible for the expenses incurred by the monitor in carrying out these tasks.

128. The names of the members of Alaska Corporate Compliance Committee, including the independent third-party environmental consultant, must be submitted for approval to the U.S. Attorney and EPA Region 10.

129. The company will then have two years to implement the Endicott EMS.

130. The purpose and scope of the audit is to evaluate the Endicott EMS to "ensure its continuing suitability, adequacy and effectiveness and to recommend any changes appropriate to achieve the objectives of the program." The third-party auditors must be certified by ANSI-RAB or have comparable credentials and experience in performing EMS audits.

port to EPA Region 10 on the status of each business unit EMS and whether it meets the objectives and elements described in the plea agreement and provide a copy of the accompanying EMS manual. Finally, the plea agreement requires that the Corporate Compliance Committees coordinate and standardize the EMS, so that a BP Amoco Group employee can transfer between business units without being forced to master different procedures.[131]

State and local governments have also utilized environmental management system commitments to settle enforcement cases. In a matter resolved in July 1999, the city of San Diego and San Diego Farah Partners and Behram Baxter, owners of a set of hotels, stipulated and consented to entry of a Judgment and Permanent Injunction, pursuant to which the hotel owners would adopt ISO 14001 in order to resolve alleged violations of the California Health and Safety Code, i.e., improper disposal of asbestos-laden floor tiles. Specifically, the defendant committed to: (i) complete an Initial ISO 14001 assessment and workplan toward completing and maintaining ISO 14001 certification for the hotels; (ii) within one year after filing of the Judgment and Permanent Injunction, complete ISO 14001 certification and forward copies of related documentation to the city; (iii) within one and one-half years of the filing of the Judgment and Permanent Injunction, hold four public meetings detailing the 14001 process and the reasons defendants undertook the program; (iv) twice a year, for three years after the filing of the Judgment and Permanent Injunction, retain an independent auditor to audit the ISO 14001 EMS, forwarding a copy of the audit report to the plaintiff; and (v) in the event the costs of the above activities total less than $92,000, pay the balance to a fund designated to support environmental enforcement.[132]

Since entering into the settlement, San Diego Farah Partners engaged the Riverdale, New Jersey, firm of FirstEnvironment to conduct an initial assessment. That assessment concluded, to the surprise of management, that there was no coherent EMS. Management had prided itself on effective and efficient management systems for other areas of their business, e.g., finance and service delivery. As a result of the assessment, management formed a committee (Committee), consisting of the general manager, head of maintenance, head of housekeeping, and the chief financial officer, to focus on environmental management.

The Committee began by preparing an EMS manual, which has been reviewed by outside experts. The reviewers found that the new manual evidences the commitment by senior management within San Diego Farah Partners to develop and implement an EMS that conforms to ISO 14001. According to counsel for the company, the attitude of senior management toward environmental awareness and regulatory compliance has improved significantly as a result of the EMS assessment. Farah Partners' employees now perceive environmental issues as an important part of their job. The company has also reduced chemical use and consumption significantly.[133]

Whereas EPA and the DOJ have tended to be nonocommittal in public statements about the use of EMS, including ISO 14001,[134] recent pronouncements are less agnostic, and may reflect deeper, broader appreciation within the federal government of the merits of EMS implementation. In *Aiming for Excellence: Actions to Encourage Stewardship and Accelerate Environmental Progress*, the Agency states: "We encourage the use of EMSs because they have the potential to improve compliance rates and environmental performance . . . It is now time for EPA to take a more prominent role in encouraging organizations to use EMSs."[135] As they wait for EPA to follow through on its pledge to focus more closely on the implications of EMS for regulation and enforcement,[136] those interested in the out-

131. Each committee must arrange for a third-party organization audit to determine if adequate coordination and standardization has occurred, with audit results to be reported directly to top management, at least once every two years.

132. State v. San Diego Farah Partners, Final Judgment in Its Entirety and Permanent Injunction Pursuant to Stipulation (Cal. App. Dep't Super. Ct. June 1999). *See Innovative Court Ruling Leads to ISO 14001 Benefits for Californian Hotel*, ISO 9000 + ISO 14000 NEWS, Mar./Apr. 2000, at 18; S. Wayne Rosenbaum, Using ISO 14001 as a Settlement Tool in Environmental Crimes Cases: A Study for the Hospitality Industry, MSI 2000, Las Vegas, Nev. (Feb. 2000); *Settling for ISO 14001?*, Int'l Envtl. Sys. Update, Aug. 1999, at 4.

133. Interview with Wayne Rosenbaum, counsel to the San Diego Farah Partners, of the law firm of Higgs, Fletcher & Mack (Jan. 26, 2000).

134. *See, e.g.*, U.S. EPA, EPA Position Statement on Environmental Management Systems and ISO 14001 and a Request for Comments on the Nature of Data to Be Collected From Environmental Management Systems/ISO 14001 Pilots, 63 Fed. Reg. 12094, 12095 (Mar. 12, 1998), wherein EPA's Office of Reinvention stated:

> Implementation of an EMS has the potential to improve an organization's environmental performance and compliance with regulatory requirements. EPA supports and will help promote the development and use of EMSs, including those based on the ISO 14001 standard, that help an organization achieve its environmental obligations and broader environmental performance goals . . . . EPA encourages the use of EMSs that focus on improved environmental performance and compliance as well as source reduction (pollution prevention) and system performance.

Id. See generally Jason Hart, *EPA Publishes Position Statement on ISO 14000*, Int'l Envtl. Sys. Update, Apr. 1998, at 1.

135. OFFICE OF THE ADMINISTRATOR, U.S. EPA, REPORT OF THE INNOVATIONS TASK FORCE, AIMING FOR EXCELLENCE: ACTIONS TO ENCOURAGE STEWARDSHIP AND ACCELERATE ENVIRONMENTAL PROGRESS 47, EPA/100-R-99-006 (July 1999) (available at <http://www.epa.gov/reinvent/taskforce/report99> and from the ELR Document Service, ELR Order No. AD-4255). *See generally EPA Report Examines Incentives, Regulatory Flexibility With EMSs*, Int'l Envtl. Sys. Update, Sept. 1999, at 1. *See also* U.S. EPA, INNOVATIVE APPROACHES TO ENFORCEMENT AND COMPLIANCE ASSURANCE: ACTION PLAN FOR INNOVATION 13, EPA/300-K-99-003 (Sept. 1999) ("[T]he enforcement and compliance assurance program will strategically promote the use of EMS in certain sectors, including sectors that are dominated by small business, to improve compliance and overall environmental performance.") (available at <http://www.epa.gov/oeca/innovative/approaches.html>). On the Clinton Administration's Executive Orders regarding the implementation of EMSs by federal agencies, *see generally Clinton Signs Executive Order for EMSs*, Int'l Envtl. Sys. Update, May 2000, at 1.

136. AIMING FOR EXCELLENCE, *supra* note 135, at 44-51. For commentary on the adaptability of EMS concepts in legislative and regulatory reform, *see generally* Jason Morrison et al., *Managing a Better Environment: Opportunities and Obstacles for ISO 14001*, *in* PUBLIC POLICY AND COMMERCE (Pacific Institute 2000); Paulette L. Stenzel, *Can the ISO 14000 Series Environmental Management Standards Provide a Viable Alternative to Government Regulation?*, 37 AM. BUS. L.J. 237 (2000); Keith Pezzoli, *Environmental Management Systems (EMSs) and Regulatory Innovation*, CAL. W. L. REV. (forthcoming 2000); *Second Generation Bill Introduced*, Int'l Envtl. Sys. Update, Dec. 1999, at 1 (reviewing H.R. 3448); Jennifer Nash & John Ehrenfeld, Environmental Management Systems and Their Role in Public Policy, Paper Presented at Environmental Protection in the New Millenium: Improving Performance, Credibility and Communication, A Summit Conference on Environmental Management Systems, Washington, D.C. (Nov. 1999); Cary Coglianese, Policy Implications of Environmental Management Systems, Paper Presented at Environmental Protection in the New Millenium: Improving Performance, Credibility and Communica-

come of that evaluation should monitor the dockets for agreements like the ones negotiated by ASARCO, BP Exploration (Alaska), and the Farah Partners. These cases and others bear witness to the growing recognition at both the state and federal levels of government that many compliance problems can be prevented, or at least promptly detected and corrected, through a well-designed EMS.[137]

tion, A Summit Conference on Environmental Management Systems, Washington, D.C. (Nov. 1999); Paul Burnet, *How Can EMSs and Regulatory Incentives Work for Everyone?*, Int'l Envtl. Sys. Update, Oct. 1999, at 12; Jason Hart, *Proposed Law to Link Regulatory Relief With Environmental Performance*, Int'l Envtl. Sys. Update, May 1999, at 1; Paula C. Murray, *Inching Toward Environmental Regulatory Reform—ISO 14000: Much Ado About Nothing or a Reinvention Tool?*, 37 AM. BUS. L.J. 35 (1999); Jay Benforado et al., *Another Parallel Track?*, ENVTL. F., Nov./Dec. 1999, at 46; David Monsma & Janice Mazurek, *Building a Performance-Based Environmental Management System: EPA's Reinvention Efforts Revisited*, 6 CORP. ENVTL. STRATEGY 233 (1999); Robert Anthony Reilly, *The New Paradigm: ISO 14000 and Its Place in Regulatory Reform*, 22 J. CORP. L. 535 (1997); Christopher L. Bell, *Why the Government Should Stay Out of ISO*, ENVTL. F., Nov./Dec. 1997, at 24; Harris Gleckman, *Promising Much But Delivering Little: ISO 14001 Should Not Be Part of Government Regulations or Procurement*, ISO 14000 Update (Bus. & the Env't), Apr. 1996, at 1.

137. This point is underscored in the statement on EMS, Compliance, and Environmental Performance, issued by the Multi-State Working Group on Environmental Management Systems on Dec. 13, 1999 ("Evidence suggests that a comprehensive approach to compliance through use of an environmental management system has merit."). *See generally MSWG Releases New Consensus Statement on ISO 14001*, Int'l Envtl. Sys. Update, Jan. 2000, at 1.

Conclusion

Federal and state governments are aware that in certain cases regulatory violations can be traced to deficient management systems. For this reason, some within the enforcement community have begun to accept reduction of a financial sanction in enforcement cases in exchange for commitments to adopt tools, techniques, and strategies that will provide greater levels of environmental protection and compliance assurance. Ovid said: "No force is greater than habit."[138] The virtues of agreements that incorporate environmental auditing, pollution prevention, and EMS elements lie in their capacity to foster practices that promise long-lasting benefits for the environment, as well as for the communities and firms involved. They are achievements worthy of praise and emulation.

138. OVID, THE ART OF LOVE 140 (Rolfe Humphries trans., Indiana Univ. Press 1957) (17 A.D.). The power of habit was well understood, as Montaigne relates, by the man who first forged the tale of the village woman who had grown used to cuddling a calf and carrying it about from the time it was born. She grew so accustomed to doing so that she was able to carry it when a fully-grown bull. Michel de Montaigne, *On Habit: And Never Easily Changing a Traditional Law*, *in* THE COMPLETE ESSAYS 122 (M.A. Screech trans., Penguin Books 1987).

APPENDIX A

STATE AUDIT PRIVILEGE AND IMMUNITY LEGISLATION AND SELF-DISCLOSURE POLICIES

STATE AUDIT PRIVILEGE AND IMMUNITY LEGISLATION AND SELF-DISCLOSURE POLICIES[1]			
Privilege Legislation	**Privilege and Immunity Legislation**	**Immunity Legislation**	**Self-Disclosure Policy**
Arkansas*	Alaska	Rhode Island	California
Illinois**	Arizona***		Connecticut
Indiana*	Colorado*		Delaware
Mississippi	Iowa		Florida
Oregon	Kansas		Indiana
	Kentucky		Maine
	Michigan*		Maryland
	Minnesota*		Massachusetts
	Montana*		Minnesota
	Nebraska		New Mexico
	Nevada		New York
	New Hampshire*		North Carolina
	Ohio*		Oregon
	South Carolina*		Pennsylvania
	South Dakota*		Tennessee
	Texas*		Vermont
	Utah*		Washington
	Virginia*		
	Wyoming*		

1. As of July 2000. Note that Idaho previously had an environmental audit privilege and immunity law that sunset on December 31, 1997.

* Denotes a state which has enacted statutory revisions and/or issued a clarifying Attorney General's statement to satisfy requirements for federally authorized, delegated or approved environmental programs.

** Denotes a state which has reached an understanding with EPA as to necessary statutory revisions and has issued a clarifying Attorney General's opinion. If amendments are enacted as agreed, the state will satisfy requirements for federally authorized, delegated or approved environmental programs.

*** On April 1, 2000, Arizona enacted a law that contains elements of audit privilege and immunity in the context of encouraging environmental management systems. The law satisfies minimum requirements for federally authorized, delegated, or approved environmental programs.

APPENDIX B

AUDIT PROVISIONS IN FEDERAL COURT CONSENT DECREES OR ORDERS

KEY:

As used in this context, the use of the terms "Yes" or "No" means that the item is addressed explicitly in the consent decree or order.

As used in this context, the use of the term "presumably" means that the item is not addressed explicitly, but can be inferred to a high degree of certainty from the language of the consent decree or order.

As used in this context, the use of the term "Not Specified" means that the item is not addressed in the consent decree or order.

As used in this context, N/A is used to abbreviate the term "Not Applicable".

As used in this table, the acronym "CEM" means continuous emissions monitor, "COM" means continuous opacity monitor, "EMS" means environmental management system, "NESHAP" means National Emission Standard for Hazardous Air Pollutants, and "SEP" means Supplemental Environmental Project.

Environmental laws addressed herein include the following: Clean Air Act ("CAA"), 42 U.S.C. §7401 *et seq.* (1998); Clean Water Act ("CWA"), 33 U.S.C. §1251 *et seq.* (1998); Comprehensive Environmental Response, Compensation, and Liability Act of 1980 ("CERCLA"), 42 U.S.C. §9601 *et seq.* (1998); Emergency Response and Community Right-To-Know Act of 1986 ("EPCRA"), 42 U.S.C. §11001 *et seq.* (1998); Endangered Species Act of 1973 ("ESA"), 16 U.S.C. §1531 *et seq.* (1998); Federal Insecticide, Fungicide, and Rodenticide Act, 7 U.S.C. §136 *et seq.* (1998); Resource Conservation and Recovery Act ("RCRA"), 42 U.S.C. §6901 *et seq.* (1998); Toxic Substances Control Act ("TSCA"), 15 U.S.C. §2601 *et seq.* (1998).

APPENDIX B, CONT'D

AUDIT PROVISIONS IN FEDERAL COURT CONSENT DECREES OR ORDERS			
Consent Decree or Order	U.S. and the Commonwealth of Pennsylvania v. A-1 Battery, Inc.	U.S. and the State of Louisiana, Dep't of Environmental Quality v. Acadia Woods Add. #2 Sewer Co., et al.	U.S. v. Air Products and Chemicals, Inc., et al.
Court	U.S. District Court, Middle District of Pennsylvania	U.S. District Court, Western District of Louisiana	U.S. District Court, District of Maryland
Date Entered	May 8, 1998	July 31, 1998	April 20, 1988
Civil Action No.	3:CV 9803 63	6:98 CV 0687	JH-88-365
Statute Violated	CERCLA	CWA, Louisiana Water Control Law	CERCLA
Type of Audit: **a) Laboratory** **b) Field** **c) Facility** **d) Compliance** **e) of an EMS**	 a) yes, if requested by EPA b) not specified c) not specified d) not specified e) not specified	 a) not specified b) not specified c) not specified d) yes e) not specified	 a) yes b) yes c) not specified d) not specified e) not specified
Frequency of Audits	not specified	not specified, but compliance monitoring by an outside auditor is required once a year for five years	not specified
Reporting Requirement as part of Audit	yes	yes	yes
Requires an EMS Provision	no	no	no
Auditing as part of EMS	N/A	N/A	N/A
Auditing as part of SEP Provision	N/A	N/A	N/A

APPENDIX B, CONT'D

AUDIT PROVISIONS IN FEDERAL COURT CONSENT DECREES OR ORDERS			
Consent Decree or Order	U.S. and Dep't of Environmental Protection v. Ajax/Acorn Manufacturing, Inc., et al.	U.S. v. Alcan Ingot and Recycling Division of the Alcan Aluminum Corp.	U.S. v. Allied-Signal, Inc.
Court	U.S. District Court, Eastern District of Pennsylvania	U.S. District Court, Western District of Kentucky, Owensboro Division	U.S. District Court, Middle District of Pennsylvania
Date Entered	October 10, 1997	May 11, 1993	July 13, 1990
Civil Action No.	89-7421	93-000-3 (CS)	CV-90-0938
Statute Violated	CERCLA	CAA	CERCLA
Type of Audit: **a) Laboratory** **b) Field** **c) Facility** **d) Compliance** **e) of an EMS**	 a) yes, if requested by EPA b) not specified c) not specified d) not specified e) not specified	 a) not specified b) not specified c) presumably yes d) presumably yes e) not specified	 a) yes b) yes c) not specified d) not specified e) not specified
Frequency of Audits	not specified	"qualitatively audit all three potlines" 4 times/wk for a period of one year; "quantitatively audit all three potlines" 1 time/wk for a period of one year	not specified
Reporting Requirement as part of Audit	yes	yes	yes
Requires an EMS	no	no	no
Auditing as part of EMS	N/A	N/A	N/A
Auditing as part of SEP Provision	N/A	N/A	N/A

APPENDIX B, CONT'D

AUDIT PROVISIONS IN FEDERAL COURT CONSENT DECREES OR ORDERS			
Consent Decree or Order	U.S. and Commonwealth of Pennsylvania v. American Color and Chemical Corp., et al.	U.S. v. American Cyanamid Co., Inc., et al. (Bush Valley Landfill)	U.S. v. American Cyanamid Co., Inc., et al. (Fike Chemical)
Court	U.S. District Court, Middle District of Pennsylvania	U.S. District Court, District of Maryland	U.S. District Court, Southern District of West Virginia
Date Entered	February 14, 1996	March 10, 1998	February 19, 1997
Civil Action No.	4: CV 92 1352	2-98-27	2: 93-0654
Statute Violated	CERCLA	CERCLA	CERCLA
Type of Audit: **a) Laboratory** **b) Field** **c) Facility** **d) Compliance** **e) of an EMS**	a) yes, if requested by EPA b) yes c) not specified d) not specified e) not specified	a) yes, if requested by EPA b) not specified c) not specified d) not specified e) not specified	a) yes, if requested by EPA b) not specified c) not specified d) not specified e) not specified
Frequency of Audits	not specified	not specified	not specified
Reporting Requirement as part of Audit	yes	yes	yes
Requires an EMS	no	no	no
Auditing as part of EMS	N/A	N/A	N/A
Auditing as part of SEP Provision	N/A	N/A	N/A

APPENDIX B, CONT'D

AUDIT PROVISIONS IN FEDERAL COURT CONSENT DECREES OR ORDERS				
Consent Decree or Order	U.S. v. Archer Daniels Midland Co.	U.S. and State of Maryland v. Edward Azrael, et al. v. Armco Inc., et al.	U.S. v. Bassett Furniture Industries, Inc.	U.S. v. Beazer East, Inc.
Court	U.S. District Court, Southern District of Iowa	U.S. District Court, District of Maryland	U.S. District Court, District of Virginia	U.S. District Court, Western District of Pennsylvania
Date Entered	September 7, 1994	June 7, 1995	October 15, 1999	October 9, 1990
Civil Action No.	4-94-CV-10436	WN-89-2898	4:99 CV 0044	90-1314
Statute Violated	CAA	CERCLA	CAA	CERCLA
Type of Audit: **a) Laboratory** **b) Field** **c) Facility** **d) Compliance** **e) of an EMS**	a) not specified b) not specified c) presumably yes d) presumably yes e) yes[1]	a) yes, if requested by EPA b) yes c) not specified d) not specified e) not specified	a) not specified b) not specified c) yes d) yes e) not specified	a) yes b) yes c) not specified d) not specified e) not specified
Frequency of Audits	not specified	not specified	"on a regular, continuing basis"	annual field audits; frequency of lab audits to be specified in a "Work Plan or other EPA-approved plan"
Reporting Requirement as part of Audit	yes	yes	yes	yes
Requires an EMS	no	no	no	no
Auditing as part of EMS	N/A	N/A	N/A	N/A
Auditing as part of SEP Provision	N/A	N/A	not specified	N/A

1. The consent decree requires a comprehensive audit of the company's environmental management practices and procedures and of how the company ensures compliance with all applicable environmental laws and regulations.

APPENDIX B, CONT'D

AUDIT PROVISIONS IN FEDERAL COURT CONSENT DECREES OR ORDERS			
Consent Decree or Order	U.S. and State of Maryland, Dep't of Environment v. Bethlehem Steel Corp.	U.S. and Commonwealth of Pennsylvania v. Joseph M. Blosenski, Jr., et al.	U.S. v. Board of County Commissioners for Cecil County, MD
Court	U.S. District Court, District of Maryland	U.S. District Court, Eastern District of Pennsylvania	U.S. District Court, District of Maryland
Date Entered	September 18, 1997	September 8, 1995	December 9, 1996
Civil Action No.	JFM-97-558, JFM-97-559	93-CV-1976	HMD-96-3082
Statute Violated	RCRA	CERCLA	CERCLA
Type of Audit: **a) Laboratory** **b) Field** **c) Facility** **d) Compliance** **e) of an EMS**	a) yes, if requested by EPA b) not specified c) not specified d) not specified e) not specified	a) yes, if requested by EPA b) yes c) not specified d) not specified e) not specified	a) yes, if requested by EPA b) yes c) not specified d) not specified e) not specified
Frequency of Audits	not specified	not specified	not specified
Reporting Requirement as part of Audit	not specified	yes	yes
Requires an EMS	no	no	no
Auditing as part of EMS	N/A	N/A	N/A
Auditing as part of SEP Provision	N/A	N/A	N/A

APPENDIX B, CONT'D

AUDIT PROVISIONS IN FEDERAL COURT CONSENT DECREES OR ORDERS			
Consent Decree or Order	U.S. v. BP America Inc., et al. (Army Creek Landfill)	U.S. and State of Louisiana v. Browning-Ferris Industries, Chemical Services, Inc., and Cecos International, Inc.	U.S. and State of Louisiana v. Browning-Ferris Industries, Chemical Services, Inc., and Cecos International, Inc.
Court	U.S. District Court, District of Delaware	U.S. District Court, Middle District of Louisiana	U.S. District Court, Western District of Louisiana, Lake Charles Division
Date Entered	September 12, 1991	October 18, 1988	August 16, 1990[2]
Civil Action No.	91-409, 91-418	87-317 Section B	88-0718-LC
Statute Violated	CERCLA	RCRA, Louisiana Environmental Quality Act[3]	RCRA, Louisiana Environmental Quality Act[4]
Type of Audit: **a) Laboratory** **b) Field** **c) Facility** **d) Compliance** **e) of an EMS**	a) yes b) yes c) not specified d) not specified e) not specified	a) not specified b) not specified c) yes d) yes e) yes	a) not specified b) not specified c) yes d) yes e) yes
Frequency of Audits	"an appropriate number" of field and lab audits	not specified	second audit one year after submission of first audit report
Reporting Requirement as part of Audit	yes	yes	yes
Requires an EMS	no	no	no
Auditing as part of EMS	N/A	yes	yes
Auditing as part of SEP Provision	N/A	N/A	N/A

2. The consent decree calls for the companies to pay a civil penalty of $1,550,000, undertake specified interim corrective actions, and institute a facility audit. An independent auditor with expertise in environmental auditing, EMSs, and environmental legal compliance, will audit facility and: (i) report on the company's environmental compliance; and (ii) analyze the company's EMSs, practices and policies at the facility and in defendants' corporate offices. The audit firm must submit proposed Audit Work Plan to EPA and Louisiana Department of Environmental Quality ("LDEQ") for approval. The auditors will not have access to attorney-client communications. Defendants must correct any violations identified in the Facility Audit Report. The Corporate Management Systems report shall: (i) identify and describe existing facility environmental operations and the corporate offices responsible for overall company-wide environmental compliance and management systems; (ii) evaluate such operations and systems, practices and policies and identify and describe perceived weaknesses; and (iii) identify and describe fully the perceived areas where defendants' EMSs may be improved as they affect the facility's compliance obligations, listing specific options for any improvements in a range of specified areas. Within 90 days of receipt of the auditor's Corporate Management Systems Report, defendants must prepare its own Corporate Management Report and Plan describing actions it has taken or will take to implement the recommendations of the independent auditors, or responding why implementations of the recommendations is not necessary. The companies must, through the use of a *different* independent audit firm, conduct a *second* environmental audit a year after submission of the Corporate Management Report and Plan. The U.S. and the State of Louisiana agreed that the Facility Audit Report and the Compliance Report and Plan could not be used as direct evidence in a proceeding to assess a civil penalty, provided the violation was corrected within specified time period.

3. Louisiana Environmental Quality Act, L.R.S. §30:1051.

4. *Id.*

APPENDIX B, CONT'D

AUDIT PROVISIONS IN FEDERAL COURT CONSENT DECREES OR ORDERS			
Consent Decree or Order	U.S. v. Burlington Northern Railroad Co., and Burlington Northern, Inc.	U.S. v. Chevron U.S.A., Inc.	U.S. v. Ciba-Geigy Corp. and Monsey Products
Court	U.S. District Court, District of Montana, Missoula Division	U.S. District Court, Eastern District of Pennsylvania	U.S. District Court, Eastern District of Pennsylvania
Date Entered	June 26, 1990	September 15, 1992	January 4, 1991
Civil Action No.	89-0-687	88-6681	91-0009
Statute Violated	CERCLA	CAA	CERCLA
Type of Audit: **a) Laboratory** **b) Field** **c) Facility** **d) Compliance** **e) of an EMS**	 a) yes b) yes c) not specified d) not specified e) not specified	 a) not specified b) not specified c) yes d) yes, for benzene e) not specified	 a) yes b) yes c) not specified d) not specified e) not specified
Frequency of Audits	"periodic" lab audits; "at least one field audit per year"	not specified	"an appropriate number"
Reporting Requirement as part of Audit	yes	yes	yes
Requires an EMS	no	no	no
Auditing as part of EMS	N/A	N/A	N/A
Auditing as part of SEP Provision	N/A	N/A	N/A

APPENDIX B, CONT'D

AUDIT PROVISIONS IN FEDERAL COURT CONSENT DECREES OR ORDERS			
Consent Decree or Order	U.S. Colorado & Eastern Railroad Co., et al.	National Wildlife Federation, et al. v. Copper Range Co.	U.S. v. Darling International, Inc.
Court	U.S. District Court, District of Colorado	U.S. District Court, Western District of Michigan, Northern Division	U.S. District Court, District of Minnesota
Date Entered	September 7, 1990	April 5, 1995	September 12, 1997
Civil Action No.	89-C-1786	2: 92-CV-186	97-1611
Statute Violated	CERCLA	CAA, CERCLA, EPCRA	CWA
Type of Audit: **a) Laboratory** **b) Field** **c) Facility** **d) Compliance** **e) of an EMS**	a) yes b) yes c) not specified d) not specified e) not specified	a) not specified b) not specified c) yes, of COM and CEM equipment only d) not specified e) not specified	a) not specified b) not specified c) yes[5] d) yes e) not specified
Frequency of Audits	"periodic" lab audits; "at least one" field audit per year	annually	not specified
Reporting Requirement as part of Audit	yes	yes	yes
Requires an EMS	no	no	no[6]
Auditing as part of EMS	N/A	N/A	yes
Auditing as part of SEP Provision	N/A	N/A	N/A

5. The decree requires a pollution prevention and waste minimization audit for the facility.

6. The decree requires the company to continue to implement its environmental compliance and management program.

APPENDIX B, CONT'D

AUDIT PROVISIONS IN FEDERAL COURT CONSENT DECREES OR ORDERS			
Consent Decree or Order	U.S. and Timothy R.E. Keeney, Commissioner of Environmental Protection of the State of Connecticut v. The Dexter Corp.	U.S. and Commonwealth of Pennsylvania v. E. I. DuPont de Nemours & Co., et al. (Bell Landfill)	U.S. v. E. I. DuPont de Nemours & Co. (Chamber Works)
Court	U.S. District Court, District of Connecticut	U.S. District Court, Middle District of Pennsylvania	U.S. District Court, District of New Jersey
Date Entered	November 13, 1992	April 2, 1997	May 21, 1991
Civil Action No.	H-89-393	3: CV 97-149	91-768
Statute Violated	CWA	CERCLA	RCRA
Type of Audit: **a) Laboratory** **b) Field** **c) Facility** **d) Compliance** **e) of an EMS**	 a) not specified b) not specified c) yes[7] d) yes e) yes	 a) yes, if requested by EPA b) yes c) not specified d) not specified e) not specified	 a) not specified b) yes c) yes d) yes e) yes
Frequency of Audits	follow-up audit required 10-12 months after first audit	not specified	not specified
Reporting Requirement as part of Audit	yes	yes	yes
Requires an EMS	no	no	no
Auditing as part of EMS	yes	N/A	yes
Auditing as part of SEP Provision	N/A	N/A	N/A

7. The decree requires a waste minimization audit for five facilities.

APPENDIX B, CONT'D

AUDIT PROVISIONS IN FEDERAL COURT CONSENT DECREES OR ORDERS			
Consent Decree or Order	U.S. v. Eagle Picher Industries, Inc.	U.S. v. Elf Atochem North America, Inc.	U.S. and State of Texas v. Encycle/Texas, Inc., and ASARCO, Inc.
Court	U.S. District Court, Western District of Missouri, Joplin Division	U.S. District Court, Eastern District of Pennsylvania	U.S. District Court, Southern District of Texas
Date Entered	September 28, 1990	May 3, 1993	October 6, 1999
Civil Action No.	87-5100-CV-SW-8	93-2182	H-99-1136
Statute Violated	CWA	CERCLA	RCRA, CWA, Texas Health and Safety and Water Codes[8]
Type of Audit: **a) Laboratory** **b) Field** **c) Facility** **d) Compliance** **e) of an EMS**	a) not specified b) not specified c) yes d) yes e) not specified	a) yes, if requested by EPA b) yes c) not specified d) not specified e) not specified	a) not specified b) not specified c) yes d) yes e) yes
Frequency of Audits	not specified	not specified	annual EMS and compliance audits for 5 years
Reporting Requirement as part of Audit	yes	yes	yes
Requires an EMS	no	no	yes
Auditing as part of EMS	N/A	N/A	yes
Auditing as part of SEP Provision	N/A	N/A	not specified

8. Chapter 361 of the Texas Health and Safety Code, Texas Health and Safety Code Ann. §§361.001 *et seq.*, and Chapter 7 of the Texas Water Code, Texas Water Code Ann. §§7.001 *et seq.*, and the rules and regulations promulgated pursuant to those statutes.

APPENDIX B, CONT'D

AUDIT PROVISIONS IN FEDERAL COURT CONSENT DECREES OR ORDERS			
Consent Decree or Order	U.S. v. Environmental International Electrical Services, Inc.	U.S. v. FMC Corp. (Pocatello Plant)	U.S. v. Lewis Frame and Ruth Frame
Court	U.S. District Court, District of Kansas	U.S. District Court, District of Idaho	U.S. District Court, Eastern District of Pennsylvania
Date Entered	April 6, 1988	July 13, 1999	August 3, 1998
Civil Action No.	88-2084-S	98-0406-E-BLW	98-CV-2844
Statute Violated	TSCA	RCRA	CERCLA
Type of Audit: **a) Laboratory** **b) Field** **c) Facility** **d) Compliance** **e) of an EMS**	a) not specified b) not specified c) yes d) yes e) not specified	a) not specified b) not specified c) yes d) yes e) yes	a) yes, if requested by EPA b) not specified c) not specified d) not specified e) not specified
Frequency of Audits	annual	currently conducts "periodic" audits; must complete a management system audit within 18 months of the date the Pocatello Plant EMS is fully implemented	not specified
Reporting Requirement as part of Audit	yes	not specified	yes
Requires an EMS	no	yes[9]	no
Auditing as part of EMS	N/A	yes	N/A
Auditing as part of SEP Provision	N/A	not specified	N/A

9. *See* discussion of EMS requirements *infra* Appendix E, note 3.

APPENDIX B, CONT'D

AUDIT PROVISIONS IN FEDERAL COURT CONSENT DECREES OR ORDERS			
Consent Decree or Order	U.S. v. GATX Corp., General American Transportation Corp.	U.S. v. Georgia-Pacific Corp.	U.S. v. Grumman St. Augustine Corp.
Court	U.S. District Court, Western District of Pennsylvania	U.S. District Court, Northern District of Georgia, Atlanta Div.ision	U.S. District Court, Middle District of Florida, Jacksonville Div.ision
Date Entered	July 3, 1995	July 16, 1996	July 20, 1993
Civil Action No.	94-312	1 96-CV-1818-FMH	91-141-CIV-J-16
Statute Violated	CERCLA	CAA	RCRA
Type of Audit: **a) Laboratory** **b) Field** **c) Facility** **d) Compliance** **e) of an EMS**	a) yes, if requested by EPA b) yes c) not specified d) not specified e) not specified	a) not specified b) not specified c) yes d) yes e) not specified	a) not specified b) not specified c) yes d) yes e) not specified
Frequency of Audits	not specified	"in accordance with G-P's normal audit cycle/schedule"	not specified
Reporting Requirement as part of Audit	yes	yes	yes
Requires an EMS	no	no	no
Auditing as part of EMS	N/A	N/A	N/A
Auditing as part of SEP Provision	N/A	N/A	N/A

APPENDIX B, CONT'D

AUDIT PROVISIONS IN FEDERAL COURT CONSENT DECREES OR ORDERS			
Consent Decree or Order	Concerned Citizens for Nuclear Safety, Inc. and Chavez v. U.S. Department of Energy and Siegfried S. Hecker	Concerned Citizens for Nuclear Safety, Inc. and Chavez v. U.S. Department of Energy and Siegfried S. Hecker (Supplemental Consent Decree)[10]	U.S. and State of Delaware v. Hercules, Inc., et al.
Court	U.S. District Court, District of New Mexico	U.S. District Court, District of New Mexico	U.S. District Court, District of Delaware
Date Entered	January 20, 1997	December 9, 1997	June 25, 1991
Civil Action No.	94-1039 M/WWD	94-1039 JP/WWD	89-CV-562-SLR
Statute Violated	CAA	CAA	CERCLA
Type of Audit: **a) Laboratory** **b) Field** **c) Facility** **d) Compliance** **e) of an EMS**	a) not specified b) not specified c) yes d) yes e) not specified	a) not specified b) not specified c) yes d) yes e) not specified	a) yes, if requested by EPA b) yes c) not specified d) not specified e) not specified
Frequency of Audits	1997, 2000, 2002, and 2003 (if necessary)	presumably 1997, 2000, 2002, and 2003 (if necessary)	not specified
Reporting Requirement as part of Audit	yes	presumably yes	yes
Requires an EMS	no	no	no
Auditing as part of EMS	N/A	N/A	N/A
Auditing as part of SEP Provision	N/A	N/A	N/A

10. The supplemental consent decree clarifies certain auditing requirements and provisions concerning coordination of plaintiff's visits to DOE's Los Alamos facility that were the subject of a dispute among the parties.

APPENDIX B, CONT'D

AUDIT PROVISIONS IN FEDERAL COURT CONSENT DECREES OR ORDERS			
Consent Decree or Order	U.S. v. Hoosier Energy Rural Electric	U.S. v. Illinois Tool Works, Inc., Slezak Enterprises, Inc.	U.S. v. John Morrell & Co. (Partial Consent Decree)
Court	U.S. District Court, Southern District of Indiana	U.S. District Court, District of Michigan, Southern Division	U.S. District Court, District of South Dakota
Date Entered	September 30, 1986	July 29, 1998	April 26, 1996
Civil Action No.	TH-85-8-C	1:98-CV-389	96-4011
Statute Violated	CAA	CERCLA	CWA
Type of Audit: **a) Laboratory** **b) Field** **c) Facility** **d) Compliance** **e) of an EMS**	 a) not specified b) not specified c) audit of CEMs d) not specified e) not specified	 a) yes b) yes c) not specified d) not specified e) not specified	 a) not specified b) not specified c) yes[11] d) not specified e) yes
Frequency of Audits	quarterly	not specified	not specified
Reporting Requirement as part of Audit	yes	presumably yes	yes
Requires an EMS	no	no	no
Auditing as part of EMS	N/A	N/A	N/A
Auditing as part of SEP Provision	N/A	N/A	N/A

11. The decree requires a pollution prevention and waste minimization audit.

APPENDIX B, CONT'D

AUDIT PROVISIONS IN FEDERAL COURT CONSENT DECREES OR ORDERS			
Consent Decree or Order	U.S. v. John Morrell & Co. (2nd Partial Consent Decree)	U.S. v. Kalama Chemical, Inc.	U.S. v. Ketchikan Pulp Co.
Court	U.S. District Court, District of South Dakota	U.S. District Court, Western District of Washington at Tacoma	U.S. District Court, District of Alaska
Date Entered	February 11, 1998	March 31, 1997	September 19, 1995
Civil Action No.	96-4011	C 95 5522 FDB	A92-587-CV (JKS)
Statute Violated	CWA	CAA	CWA, CAA
Type of Audit: **a) Laboratory** **b) Field** **c) Facility** **d) Compliance** **e) of an EMS**	a) not specified b) not specified c) yes d) yes e) yes	a) not specified b) not specified c) yes[12] d) not specified e) not specified	a) not specified b) not specified c) yes d) yes e) yes
Frequency of Audits	not specified	not specified	not specified
Reporting Requirement as part of Audit	yes	yes	yes
Requires an EMS	yes[13]	no	no
Auditing as part of EMS	presumably yes	N/A	not specified
Auditing as part of SEP Provision	N/A	yes	N/A

12. The SEP provision of the decree requires a pollution prevention audit and a fugitive emissions audit of defendant's facility.

13. *See* discussion of EMS requirements *infra* Appendix E, note 10.

APPENDIX B, CONT'D

AUDIT PROVISIONS IN FEDERAL COURT CONSENT DECREES OR ORDERS			
Consent Decree or Order	U. S. v. Kowinsky Farms, Inc., et al.	U.S. v. Lord Corporation (Lord Shope Superfund Site)	U.S. v. Lord Corporation (Saegertown Industrial Area)
Court	U.S. District Court, District of Delaware	U.S. District Court, Western District of Pennsylvania	U.S. District Court, Western District of Pennsylvania
Date Entered	April 8, 1992	September 27, 1991	March 15, 1994
Civil Action No.	92-04-JJF	91-177E	94-43
Statute Violated	CERCLA	CERCLA	CERCLA
Type of Audit: **a) Laboratory** **b) Field** **c) Facility** **d) Compliance** **e) of an EMS**	 a) yes, if requested by EPA b) yes c) not specified d) not specified e) not specified	 a) yes b) yes c) not specified d) not specified e) not specified	 a) yes, if requested by EPA b) yes c) not specified d) not specified e) not specified
Frequency of Audits	not specified	"an appropriate number"	not specified
Reporting Requirement as part of Audit	yes	yes	yes
Requires an EMS	no	no	no
Auditing as part of EMS	N/A	N/A	N/A
Auditing as part of SEP Provision	N/A	N/A	N/A

APPENDIX B, CONT'D

AUDIT PROVISIONS IN FEDERAL COURT CONSENT DECREES OR ORDERS				
Consent Decree or Order	U.S. and State of Idaho v. The M.A. Hanna Co., et al.	U.S. v. Marine Shale Processors, Inc., and Recycling Park, Inc., et al.	U.S. v. Meadow Gold Dairies, Inc.	U.S. v. Menominee Paper Co., and Bell Packaging Corp.
Court	U.S. District Court, District of Idaho	U.S. District Court, Western District of Louisiana, Lafayette Division	U.S. District Court, District of Montana, Helena Division	U.S. District Court, Western District of Michigan, Northern Division
Date Entered	April 28, 1995	February 20, 1998	Oct. 21, 1993	July 20, 1990
Civil Action No.	83-4179 (R)	CV90-1240	CV 91-15-H-CCL	88-108-2
Statute Violated	CERCLA, CWA, ESA, Idaho Environmental Protection and Health Act[14]	RCRA, CWA	CWA	CWA
Type of Audit: **a) Laboratory** **b) Field** **c) Facility** **d) Compliance** **e) of an EMS**	a) yes, if requested by Trustees b) yes, if requested by Trustees c) not specified d) not specified e) not specified	a) not specified b) not specified c) yes[15] d) not specified e) not specified	a) not specified b) not specified c) yes d) yes e) not specified	a) not specified b) not specified c) yes d) yes e) not specified
Frequency of Audits	up to one lab audit and field audit per year, if requested	annually	not specified	not specified
Reporting Requirement as part of Audit	yes, if audit is requested	yes	yes	yes
Requires an EMS	no	no	no	no
Auditing as part of EMS	N/A	N/A	N/A	N/A
Auditing as part of SEP Provision	N/A	N/A	N/A	N/A

14. Idaho Environmental Protection and Health Act, Idaho Code §§39-101 *et seq.*

15. The decree requires an annual waste management audit.

APPENDIX B, CONT'D

AUDIT PROVISIONS IN FEDERAL COURT CONSENT DECREES OR ORDERS			
Consent Decree or Order	U.S. v. Modern Trash Removal of York, Inc.	U.S. and State of Michigan, et al. v. Monitor Sugar Co.	U.S. v. Nassau Metal Corp., Inc., et al.
Court	U.S. District Court, Middle District of Pennsylvania	U.S. District Court, Eastern District of Michigan, Northern Division	U.S. District Court, Middle District of Pennsylvania
Date Entered	June 14, 1993	December 10, 1987	July 22, 1998
Civil Action No.	92-0819	85-10309 (CV-BC)	3:96-CV-562
Statute Violated	CERCLA	CAA	CERCLA
Type of Audit: **a) Laboratory** **b) Field** **c) Facility** **d) Compliance** **e) of an EMS**	a) yes, if requested by EPA b) presumably no c) not specified d) not specified e) not specified	a) not specified b) not specified c) COM audit only d) not specified e) not specified	a) yes, if requested by EPA b) yes c) not specified d) not specified e) not specified
Frequency of Audits	not specified	not specified	not specified
Reporting Requirement as part of Audit	yes	yes	yes
Requires an EMS	no	no	no
Auditing as part of EMS	N/A	N/A	N/A
Auditing as part of SEP Provision	N/A	N/A	N/A

APPENDIX B, CONT'D

AUDIT PROVISIONS IN FEDERAL COURT CONSENT DECREES OR ORDERS			
Consent Decree or Order	U.S. v. Neville Land Co. and Wilmington Securities, Inc.	U.S. v. New Haven Foundry, Inc.	U.S. v. O'Hara, William J., et al.
Court	U.S. District Court, Western District of Pennsylvania	U.S. District Court, Eastern District of Michigan	U.S. District Court, Eastern District of Pennsylvania
Date Entered	December 31, 1997	June 23, 1997	May 10, 1989
Civil Action No.	97-1683	97-71842	89-1383
Statute Violated	CERCLA	RCRA	CERCLA
Type of Audit: **a) Laboratory** **b) Field** **c) Facility** **d) Compliance** **e) of an EMS**	a) yes, if requested by EPA b) not specified c) not specified d) not specified e) not specified	a) yes b) not specified c) yes d) yes e) not specified	a) yes b) yes c) not specified d) not specified e) not specified
Frequency of Audits	not specified	not specified	"an appropriate number"
Reporting Requirement as part of Audit	yes	yes	yes
Requires an EMS	no	no	no
Auditing as part of EMS	N/A	N/A	N/A
Auditing as part of SEP Provision	N/A	N/A	N/A

APPENDIX B, CONT'D

AUDIT PROVISIONS IN FEDERAL COURT CONSENT DECREES OR ORDERS			
Consent Decree or Order	U.S. v. Occidental Chemical Corporation, Bridgestone/Firestone, Inc. (Occidental Chemical Superfund Site)[16]	U.S. v. Occidental Chemical Corp. (Pottstown Plant)	U.S. v. Olin Corporation
Court	U.S. District Court, Eastern District of Pennsylvania	U.S. District Court, Eastern District of Pennsylvania	U.S. District Court, Western District of Virginia
Date Entered	June 19, 1997	August 11, 1990	July 29, 1997
Civil Action No.	96-CV-6558	89-5687	97-0090-A
Statute Violated	CERCLA	CAA	CERCLA
Type of Audit: **a) Laboratory** **b) Field** **c) Facility** **d) Compliance** **e) of an EMS**	a) yes, if requested by EPA b) yes c) not specified d) not specified e) not specified	a) not specified b) not specified c) yes d) yes, as to the vinyl chloride NESHAP only e) not specified	a) yes b) not specified c) not specified d) not specified e) not specified
Frequency of Audits	not specified	every 14 months	not specified
Reporting Requirement as part of Audit	yes	yes	yes
Requires an EMS	no	no	no
Auditing as part of EMS	N/A	N/A	N/A
Auditing as part of SEP Provision	N/A	N/A	N/A

16. This consent decree incorporates by reference EPA Region III administrative order No. III-94-26-DC, issued June 23, 1994.

APPENDIX B, CONT'D

AUDIT PROVISIONS IN FEDERAL COURT CONSENT DECREES OR ORDERS			
Consent Decree or Order	U.S. v. Pennsylvania Power & Light Co. and Union Gas Co.	U.S. v. Marvin Pesses, et al.	U.S. v. Pfaltz & Bauer, Inc.
Court	U.S. District Court, Eastern District of Pennsylvania	U.S. District Court, Western District of Pennsylvania	U.S. District Court, District of Connecticut
Date Entered	September 1, 1992	June 19, 1997	December 16, 1996
Civil Action No.	92-CV-2730	90-0654	396 CV 00305 PCD
Statute Violated	CERCLA	CERCLA	RCRA, CERCLA
Type of Audit: **a) Laboratory** **b) Field** **c) Facility** **d) Compliance** **e) of an EMS**	 a) yes, if requested by EPA b) yes c) not specified d) not specified e) not specified	 a) yes, if requested by EPA b) not specified c) not specified d) not specified e) not specified	 a) yes b) not specified c) yes d) yes e) not specified
Frequency of Audits	not specified	not specified	not specified
Reporting Requirement as part of Audit	yes	yes	yes
Requires an EMS	no	no	no
Auditing as part of EMS	N/A	N/A	N/A
Auditing as part of SEP Provision	N/A	N/A	N/A

APPENDIX B, CONT'D

AUDIT PROVISIONS IN FEDERAL COURT CONSENT DECREES OR ORDERS			
Consent Decree or Order	U.S. v. Pizzuto Company, Inc., and Reynolds Metals Co., Inc.	U.S. v. Pneumo Abex Corp., et al.	U.S. v. PPG Industries, Inc., and Aluminum Company of America
Court	U.S. District Court, Western District of Arkansas, Hot Springs Division	U.S. District Court, Eastern District of Virginia, Norfolk Division	U.S. District Court, Western District of Pennsylvania
Date Entered	November 28, 1989	April 25, 1996	October 24, 1991
Civil Action No.	88-6097	2:96-CV-27	91-1276
Statute Violated	CAA	CERCLA	CERCLA
Type of Audit: **a) Laboratory** **b) Field** **c) Facility** **d) Compliance** **e) of an EMS**	a) not specified b) not specified c) yes, asbestos only d) yes[17] e) not specified	a) yes, if requested by EPA b) yes c) not specified d) not specified e) not specified	a) yes b) yes c) not specified d) not specified e) not specified
Frequency of Audits	not specified	not specified	"an appropriate number"[18]
Reporting Requirement as part of Audit	yes	yes	yes
Requires an EMS	no	no	no
Auditing as part of EMS	N/A	N/A	N/A
Auditing as part of SEP Provision	N/A	N/A	N/A

17. The decree requires that in connection with all asbestos abatement projects subject to the CAA asbestos NESHAP at Reynolds facilities in Arkansas, Reynolds shall use a specified audit procedure for each contractor to insure that the contractor's work is in compliance with the asbestos NESHAP.

18. The frequency of lab audits is to be specified in a work plan.

APPENDIX B, CONT'D

AUDIT PROVISIONS IN FEDERAL COURT CONSENT DECREES OR ORDERS			
Consent Decree or Order	U.S. v. Raymark Corporation, et al.	U.S. and State of Indiana v. Reilly Industries, Inc.	U.S. v. Reynolds Metals Co. and Westvaco Corp.
Court	U.S. District Court, Eastern District of Pennsylvania	U.S. District Court, Southern District of Indiana, Indianapolis Division	U.S. District Court, Eastern District of Virginia
Date Entered	February 21, 1989	September 19, 1998	November 5, 1997
Civil Action No.	85-3073	5:98 CV 1409	3:97-CV-226
Statute Violated	RCRA	CERCLA	CERCLA
Type of Audit: **a) Laboratory** **b) Field** **c) Facility** **d) Compliance** **e) of an EMS**	 a) yes, if requested by EPA b) yes c) not specified d) not specified e) not specified	 a) yes b) yes c) not specified d) not specified e) not specified	 a) yes, if requested by EPA b) not specified c) not specified d) not specified e) not specified
Frequency of Audits	not specified	not specified	not specified
Reporting Requirement as part of Audit	presumably yes	presumably yes	yes
Requires an EMS	no	no	no
Auditing as part of EMS	N/A	N/A	N/A
Auditing as part of SEP Provision	N/A	N/A	N/A

APPENDIX B, CONT'D

AUDIT PROVISIONS IN FEDERAL COURT CONSENT DECREES OR ORDERS			
Consent Decree or Order	U.S. v. Rhone-Poulenc Ag. Co., and Union Carbide Chemical and Plastics Co., Inc.	U.S. v. Rhone-Poulenc Basic Chemicals (Stauffer Chem. Co.)	U. S. v. Ralph R. Riehl, Jr., et al. v. Parker White Metal Co., et al., v. Klausner Barrel Co.
Court	U.S. District Court, Southern District of West Virginia	U.S. District Court, District of Montana, Butte Division	U.S. District Court, Western District of Pennsylvania, Erie Division
Date Entered	December 19, 1991	March 28, 1989	April 26, 1996
Civil Action No.	2:90-0929	CV-89-17-BU	89-226 (Erie)
Statute Violated	CWA	CAA	CERCLA
Type of Audit: **a) Laboratory** **b) Field** **c) Facility** **d) Compliance** **e) of an EMS**	a) not specified b) not specified c) yes d) yes e) yes	a) not specified b) not specified c) of each COM only d) not specified e) not specified	a) yes, if requested by EPA b) yes c) not specified d) not specified e) not specified
Frequency of Audits	not specified	semi-annual	not specified
Reporting Requirement as part of Audit	yes	yes	yes
Requires an EMS	no	no	no
Auditing as part of EMS	not specified	N/A	N/A
Auditing as part of SEP Provision	N/A	N/A	N/A

APPENDIX B, CONT'D

AUDIT PROVISIONS IN FEDERAL COURT CONSENT DECREES OR ORDERS			
Consent Decree or Order	U.S. v. Rohm & Haas Co., and Smithkline Beecham Corp.	U.S. v. Scovill, Inc.	U.S. v. Harold Shane v. A&B Iron and Metal Co. v. Ace Battery Co., Inc. (Arcanum Iron & Metal Superfund Site)
Court	U.S. District Court, Middle District of Pennsylvania	U.S. District Court, Eastern District of Virginia, Norfolk Division	U.S. District Court, Southern District of Ohio, Western Division
Date Entered	February 9, 1993	May 18, 1995	March 20, 1998
Civil Action No.	92-1295	3: 95-CV-159	C-3-89-383; 90-0102-C
Statute Violated	CERCLA	CERCLA	CERCLA
Type of Audit: **a) Laboratory** **b) Field** **c) Facility** **d) Compliance** **e) of an EMS**	a) yes, if requested by EPA b) yes c) not specified d) not specified e) not specified	a) yes, if requested by EPA b) yes c) not specified d) not specified e) not specified	a) yes b) yes c) not specified d) not specified e) not specified
Frequency of Audits	not specified	not specified	not specified
Reporting Requirement as part of Audit	yes	yes	yes
Requires an EMS	no	no	no
Auditing as part of EMS	N/A	N/A	N/A
Auditing as part of SEP Provision	N/A	N/A	N/A

APPENDIX B, CONT'D

AUDIT PROVISIONS IN FEDERAL COURT CONSENT DECREES OR ORDERS			
Consent Decree or Order	U.S. v. Terry L. Shaner, et al.	U.S. v. Southeastern Pennsylvania Transportation Authority, et al.	U.S. and State of Washington, Dep't of Ecology v. Tacoma, City of
Court	U.S. District Court, Eastern District of Pennsylvania	U.S. District Court, Eastern District of Pennsylvania	U.S. District Court, Western District of Washington
Date Entered	January 26, 1995	February 25, 1986	June 14, 1994
Civil Action No.	85-1372	86-1094	C94-5193
Statute Violated	CERCLA	CERCLA	CWA
Type of Audit: **a) Laboratory** **b) Field** **c) Facility** **d) Compliance** **e) of an EMS**	a) yes, if requested by EPA b) yes c) not specified d) not specified e) not specified	a) yes b) yes c) not specified d) not specified e) not specified	a) not specified b) not specified c) yes d) yes e) not specified
Frequency of Audits	not specified	not specified	every 6 months after the required plant improvements
Reporting Requirement as part of Audit	yes	yes	yes
Requires an EMS	no	no	no
Auditing as part of EMS	N/A	N/A	N/A
Auditing as part of SEP Provision	N/A	N/A	N/A

APPENDIX B, CONT'D

AUDIT PROVISIONS IN FEDERAL COURT CONSENT DECREES OR ORDERS			
Consent Decree or Order	U.S. v. Temrac Inc., and Sunbeam-Oster Co., Inc.	U.S. v. Texaco Refining & Marketing, Inc., f/k/a Getty Refining and Marketing Co., and Texaco Chemical Co.	U.S. v. Trident Seafoods Corp.
Court	U.S. District Court, Eastern District of Pennsylvania	U.S. District Court, District of Delaware	U.S. District Court, District of Alaska
Date Entered	July 18, 1991	September 1, 1988	February 24, 1998
Civil Action No.	91-3043	86-321-MMS	A97-0093CV (HRH)
Statute Violated	CERCLA	CAA	CWA
Type of Audit: **a) Laboratory** **b) Field** **c) Facility** **d) Compliance** **e) of an EMS**	a) yes b) yes c) not specified d) not specified e) not specified	a) not specified b) not specified c) yes, a "benzene" audit of the refinery d) yes e) not specified	a) not specified b) not specified c) yes d) yes e) yes
Frequency of Audits	an "appropriate number" of lab audits at a frequency specified in the work plan; an "appropriate number" of field audits.	not specified	audits every other year
Reporting Requirement as part of Audit	yes	yes	yes
Requires an EMS	no	no	yes[19]
Auditing as part of EMS	N/A	N/A	yes
Auditing as part of SEP Provision	N/A	N/A	N/A

19. *See* discussion of EMS requirements *infra* Appendix E, note 11.

APPENDIX B, CONT'D

AUDIT PROVISIONS IN FEDERAL COURT CONSENT DECREES OR ORDERS			
Consent Decree or Order	U.S. v. Trustees of Boston University	U.S. v. TurboCombustor Technology, Inc.	U.S. v. Tyson, Franklin P., Ciba-Geigy Corp. and General Devices, Inc. v. Wyeth Laboratories, Inc., et al.
Court	U.S. District Court, District of Massachusetts	U.S. District Court, Southern District of Florida	U.S. District Court, Eastern District of Pennsylvania
Date Entered	December 9, 1997	September 24, 1997	June 21, 1988
Civil Action No.	97-12261-PBS	97-14274-CIV-KLR	84-2663
Statute Violated	CWA, RCRA	CWA	CERCLA
Type of Audit: **a) Laboratory** **b) Field** **c) Facility** **d) Compliance** **e) of an EMS**	a) not specified b) not specified c) yes d) yes e) not specified	a) not specified b) not specified c) yes d) yes e) not specified	a) yes b) yes c) not specified d) not specified e) not specified
Frequency of Audits	not specified	annually	not specified
Reporting Requirement as part of Audit	yes	yes	yes
Requires an EMS	no[20]	no	no
Auditing as part of EMS	yes	N/A	N/A
Auditing as part of SEP Provision	yes	N/A	N/A

20. The Consent Decree does, however, contain a requirement that Boston University submit a report to EPA describing the University's environmental management system, analyzing the ability of the system to prevent noncompliance and to the extent warranted, recommending measures to improve the system and the ability of the system to ensure that environmental laws are not violated.

APPENDIX B, CONT'D

AUDIT PROVISIONS IN FEDERAL COURT CONSENT DECREES OR ORDERS			
Consent Decree or Order	U.S. and Timothy R.E. Keeney, Commissioner of Environmental Protection of the State of Connecticut v. United Technologies, Inc.	U.S. v. United Technologies Automotive, Inc.	U.S. v. USX Corporation (Gary Works)
Court	U.S. District Court, District of Connecticut	U.S. District Court, Southern District of Ohio, Eastern Division	U.S. District Court, Northern District of Indiana
Date Entered	August 23, 1993	December 11, 1992	January 26, 1999
Civil Action No.	H-90-715 (JAC)	2-92-795	2: 98-CV-465 JM
Statute Violated	RCRA, CWA, Connecticut Water Pollution Control Act	CERCLA	CWA
Type of Audit: **a) Laboratory** **b) Field** **c) Facility** **d) Compliance** **e) of an EMS**	a) not specified b) not specified c) yes d) yes e) yes	a) yes b) yes c) not specified d) not specified e) not specified	a) not specified b) not specified c) yes d) yes e) yes
Frequency of Audits	annually for three years	not specified	annually
Reporting Requirement as part of Audit	yes	presumably yes	yes
Requires an EMS	yes[21]	no	yes[22]
Auditing as part of EMS	yes	N/A	yes
Auditing as part of SEP Provision	N/A	N/A	N/A

21. *See* discussion of EMS requirements *infra* Appendix E, note 19.

22. *Id.*, Appendix E, note 20.

APPENDIX B, CONT'D

AUDIT PROVISIONS IN FEDERAL COURT CONSENT DECREES OR ORDERS			
Consent Decree or Order	U.S. v. USX Corporation (Yeoman Creek Landfill Superfund Site)	Natural Resources Defense Counsel, San Diego Baykeeper, Kenneth J. Moser v. James W. Van Loben Sels and U.S. v. California Dept. of Transportation	U.S. v. Virginia Properties, Inc.
Court	U.S. District Court, Northern District of Illinois, Eastern Division	U.S. District Court, Southern District of California	U.S. District Court, Eastern District of Virginia
Date Entered	February 4, 1999	March 10, 1998	September 30, 1994
Civil Action No.	98 C 6389	96-1440-IEG (POR), 97-0037-IEG (POR)	3:94-CV498
Statute Violated	CERCLA	CWA	CERCLA
Type of Audit: **a) Laboratory** **b) Field** **c) Facility** **d) Compliance** **e) of an EMS**	 a) yes b) yes c) not specified d) not specified e) not specified	 a) not specified b) not specified c) yes d) yes e) not specified	 a) yes, if requested by EPA b) yes c) not specified d) not specified e) not specified
Frequency of Audits	not specified	annually	not specified
Reporting Requirement as part of Audit	presumably yes	yes	yes
Requires an EMS	no	no	no
Auditing as part of EMS	N/A	N/A	N/A
Auditing as part of SEP Provision	N/A	N/A	N/A

APPENDIX B, CONT'D

AUDIT PROVISIONS IN FEDERAL COURT CONSENT DECREES OR ORDERS			
Consent Decree or Order	U.S. v. Weyerhaeuser Company	U.S. and Natural Resources Defense Council, Inc. v. Wheeling-Pittsburgh Steel Corp. (Allenport facility)	U.S. v. Wheeling-Pittsburgh Steel Corp. (Steubenville, Mingo Junction, and Yorkville facilities)
Court	U.S. District Court, Western District of Wisconsin	U.S. District Court, Western District of Pennsylvania	U.S. District Court, Southern District of Ohio, Eastern Division
Date Entered	January 3, 1990	November 4, 1992	July 16, 1991
Civil Action No.	89-C-0973-C	89-237 S	C288-598
Statute Violated	CAA	CWA	CWA
Type of Audit: **a) Laboratory** **b) Field** **c) Facility** **d) Compliance** **e) of an EMS**	 a) not specified b) not specified c) yes, of CEM only d) not specified e) not specified	 a) not specified b) not specified c) yes d) yes e) yes	 a) not specified b) not specified c) yes d) yes e) yes
Frequency of Audits	not specified	every other year	every other year
Reporting Requirement as part of Audit	yes	yes	yes
Requires an EMS	no	yes[23]	no[24]
Auditing as part of EMS	N/A	yes	yes
Auditing as part of SEP Provision	N/A	N/A	N/A

23. *See* discussion of EMS requirements *infra* note 24.

24. Wheeling-Pittsburgh Steel Corp. ("WPSC") entered into a Consent Decree with U.S. EPA on November 4, 1992 in order to resolve allegations that it had violated the Clean Water Act by discharging pollutants from a facility in Allenport, Pennsylvania, in excess of permitted levels. As part of its agreement with the government, WPSC committed to retain an outside audit firm to conduct an environmental management system audit. The company was also required to submit a plan to the Agency describing the actions it intended to take to implement the EMS recommendations of the independent auditor. *See* United States v. Wheeling-Pittsburgh Steel Corp., No. 89-237 S (W.D. Pa. Nov. 4, 1992) (Consent Decree).

APPENDIX B, CONT'D

AUDIT PROVISIONS IN FEDERAL COURT CONSENT DECREES OR ORDERS			
Consent Decree or Order	U.S. v. Witco Corporation (The Halloy Chemical Superfund Site)	U.S. v. Witco Corporation (Trisa Spill Site)	U.S. v Yaffe Iron and Metal Co., Inc.
Court	U.S. District Court, District of Delaware	U.S. District Court, District of Delaware	U.S. District Court, Eastern District of Oklahoma
Date Entered	April 9, 1992	April 19, 1991	February 26, 1997
Civil Action No.	92-93	91-022	Civ-95-308-B
Statute Violated	CERCLA	CERCLA	CWA
Type of Audit: **a) Laboratory** **b) Field** **c) Facility** **d) Compliance** **e) of an EMS**	a) yes, if requested by EPA b) yes c) not specified d) not specified e) not specified	a) yes b) yes c) not specified d) not specified e) not specified	a) not specified b) not specified c) yes d) yes e) not specified
Frequency of Audits	not specified	"an appropriate number"	only one audit required
Reporting Requirement as part of Audit	yes	yes	yes
Requires an EMS	no	no	no
Auditing as part of EMS	N/A	N/A	N/A
Auditing as part of SEP Provision	N/A	N/A	yes

APPENDIX C

AUDIT PROVISIONS IN EPA ADMINISTRATIVE ORDERS

KEY:

As used in this context, the use of the terms "Yes" or "No" means that the item is addressed explicitly in the administrative order.

As used in this context, the use of the term "presumably" means that the item is not addressed explicitly, but can be inferred to a high degree of certainty from the language of the administrative order.

As used in this context, the use of the term "Not Specified" means that the item is not addressed in the administrative order.

As used in this context, N/A is used to abbreviate the term "Not Applicable".

As used in this table, the acronym "CEM" means continuous emissions monitor, "COM" means continuous opacity monitor, "EMS" means environmental management system, "NESHAP" means National Emission Standard for Hazardous Air Pollutants, and "SEP" means Supplemental Environmental Project.

Environmental laws addressed herein include the following: Clean Air Act ("CAA"), 42 U.S.C. §7401 *et seq.* (1998); Clean Water Act ("CWA"), 33 U.S.C. §1251 *et seq.* (1998); Comprehensive Environmental Response, Compensation, and Liability Act of 1980 ("CERCLA"), 42 U.S.C. §9601 *et seq.* (1998); Emergency Response and Community Right-To-Know Act of 1986 ("EPCRA"), 42 U.S.C. §11001 *et seq.* (1998); Endangered Species Act of 1973 ("ESA"), 16 U.S.C. §1531 *et seq.* (1998); Federal Insecticide, Fungicide, and Rodenticide Act, 7 U.S.C. §136 *et seq.* (1998); Resource Conservation and Recovery Act ("RCRA"), 42 U.S.C. §6901 *et seq.* (1998); Toxic Substances Control Act ("TSCA"), 15 U.S.C. §2601 *et seq.* (1998).

APPENDIX C, CONT'D

AUDIT PROVISIONS IN EPA ADMINISTRATIVE ORDERS			
Order	In re Chemical Waste Management, Inc.	Chem-Security Systems, Inc.	In the Matter of Formosa Plastics Corp.
Date Entered	December 19, 1984	July 18, 1985	February 27, 1991
Docket No.	TSCA-84-H-03	1085-06-08-3008P	VI-010-H
Statute Violated	TSCA, RCRA	RCRA	RCRA
Type of Audit: **a) Laboratory** **b) Field** **c) Facility** **d) Compliance** **e) of an EMS**	a) not specified b) not specified c) yes d) yes e) yes	a) not specified b) not specified c) yes d) yes e) not specified	a) not specified b) not specified c) yes d) yes e) not specified
Frequency of Audits	not specified	quarterly audits for one year	annually for four years
Reporting Requirement as part of Audit	yes	yes	yes
Requires an EMS	no	no[1]	no
Auditing as part of EMS	yes	not specified	N/A
Auditing as part of SEP Provision	N/A	N/A	N/A

1. The language in this consent order represents a "transitional" period when environmental management systems were first being developed. Even though the order itself does not require an EMS, the position descriptions of both the Environmental Compliance Coordinator and the Environmental Compliance Officer (Exhibit B, 6 and 8, respectively) incorporated into the order require implementation and operation of an EMS.

APPENDIX C, CONT'D

AUDIT PROVISIONS IN EPA ADMINISTRATIVE ORDERS			
Order	In the Matter of Halocarbon Products Corp.	Hydrolabs, Inc.	Little America Refining Co.
Date Entered	August 2, 1993	December 8, 1994	October 31, 1994
Docket No.	TSCA-90-H-18	TSCA-94-H-03	TSCA-93-H-09
Statute Violated	TSCA	TSCA	TSCA
Type of Audit: **a) Laboratory** **b) Field** **c) Facility** **d) Compliance** **e) of an EMS**	a) not specified b) not specified c) presumably yes d) yes e) not specified	a) not specified b) not specified c) yes d) yes e) not specified	a) not specified b) not specified c) yes, several refineries d) yes e) not specified
Frequency of Audits	"an audit"	only one audit required	only one audit required
Reporting Requirement as part of Audit	yes	yes	yes
Requires an EMS	no	no	no
Auditing as part of EMS	N/A	N/A	N/A
Auditing as part of SEP Provision	N/A	yes	yes

APPENDIX C, CONT'D

AUDIT PROVISIONS IN EPA ADMINISTRATIVE ORDERS			
Order	Marmab, Inc.	In the Matter of Monsanto Company	In the Matter of Novak Sanitary Landfill Site, et al.[2]
Date Entered	December 8, 1994	January 3, 1990	June 30, 1995
Docket No.	TSCA-94-H-15	TSCA 89-H-21	III-95-52-DC
Statute Violated	TSCA	TSCA	CERCLA
Type of Audit: **a) Laboratory** **b) Field** **c) Facility** **d) Compliance** **e) of an EMS**	a) not specified b) not specified c) yes d) yes e) not specified	a) not specified b) not specified c) presumably yes d) yes e) not specified	a) yes b) yes c) not specified d) not specified e) not specified
Frequency of Audits	only one audit required	"an audit"	"an appropriate number" of lab audits; "at least one" field audit
Reporting Requirement as part of Audit	yes	yes	yes
Requires an EMS	no	no	no
Auditing as part of EMS	N/A	N/A	N/A
Auditing as part of SEP Provision	yes	N/A	N/A

2. This administrative order is incorporated by reference into and made part of a federal court consent decree in United States v. Air Products and Chem., Inc., C.V. No. 97-CV-0674 (E.D. Pa., order entered December 30, 1998), but the federal decree itself does not require the audit.

APPENDIX C, CONT'D

AUDIT PROVISIONS IN EPA ADMINISTRATIVE ORDERS			
Order	In the Matter of the P.D. George Company	In the Matter of Rubber Engineering, Inc., a Division of Baker Hughes, Inc.	In the Matter of Sandoz Crop Protection Corp.
Date Entered	October 2, 1990	September 6, 1989	April 3, 1991
Docket No.	TSCA-89-H-05	EPCRA-VIII-89-01	FIFRA-90-H-7
Statute Violated	TSCA	EPCRA	FIFRA
Type of Audit: **a) Lab** **b) Field** **c) Facility** **d) Compliance** **e) of an EMS**	a) not specified b) not specified c) presumably yes d) yes[3] e) not specified	a) not specified b) not specified c) yes[4] d) presumably yes e) not specified	a) not specified b) not specified c) not specified d) yes[5] e) not specified
Frequency of Audits	"an internal audit"	"an audit"	not specified
Reporting Requirement as part of Audit	yes	yes	yes
Requires an EMS	no	no	no
Auditing as part of EMS Program	N/A	N/A	N/A
Auditing as part of SEP Provision	N/A	N/A	N/A

3. PDG agrees to conduct an internal audit to review and report on its compliance with TSCA §§5 and 8.

4. The decree requires Rubber Engineering, Inc. to "develop and implement" an audit of environmental pollutants at its Salt Lake City facility.

5. The audit is to determine whether the pesticide products that Sandoz exports comply with FIFRA §17(a)(2).

APPENDIX C, CONT'D

AUDIT PROVISIONS IN EPA ADMINISTRATIVE ORDERS			
Order	Sanncor Industries, Inc.	Sara Lee Corp.	Woodlawn Landfill Site, Cecil County
Date Entered	April 22, 1993	November 4, 1993	November 25, 1994
Docket No.	TSCA-92-H-07	EPCRA/CERCLA H-93-01	III-95-05-DC
Statute Violated	TSCA	EPCRA, CERCLA	CERCLA
Type of Audit: **a) Lab** **b) Field** **c) Facility** **d) Compliance** **e) of an EMS**	 a) not specified b) not specified c) yes d) yes e) not specified	 a) not specified b) not specified c) yes d) yes e) not specified	 a) yes b) yes c) not specified d) not specified e) not specified
Frequency of Audits	only one audit required	only one audit of multiple facilities required	"an appropriate number" of lab audits; "at least one" field audit
Reporting Requirement as part of Audit	yes	yes	yes
Requires an EMS	no	no	no
Auditing as part of EMS Program	N/A	N/A	N/A
Auditing as part of SEP Provision	yes	yes	N/A

APPENDIX D

POLLUTION PREVENTION SUPPLEMENTAL ENVIRONMENTAL PROJECTS IN FISCAL YEAR 1999

KEY:

Environmental laws addressed herein include the following: Clean Air Act ("CAA"), 42 U.S.C. §7401 *et seq.* (1998); Clean Water Act ("CWA"), 33 U.S.C. §1251 *et seq.* (1998); Comprehensive Environmental Response, Compensation, and Liability Act of 1980 ("CERCLA"), 42 U.S.C. §9601 *et seq.* (1998); Emergency Response and Community Right-To-Know Act of 1986 ("EPCRA"), 42 U.S.C. §11001 *et seq.* (1998); Endangered Species Act of 1973 ("ESA"), 16 U.S.C. §1531 *et seq.* (1998); Federal Insecticide, Fungicide, and Rodenticide Act, 7 U.S.C. §136 *et seq.* (1998); Resource Conservation and Recovery Act ("RCRA"), 42 U.S.C. §6901 *et seq.* (1998); Toxic Substances Control Act ("TSCA"), 15 U.S.C. §2601 *et seq.* (1998).

POLLUTION PREVENTION SUPPLEMENTAL ENVIRONMENTAL PROJECTS IN FISCAL YEAR 1999			
Order	A-E Farms, Inc.	Ajinomoto USA	Allied Universal
Final Order Entered	February 5, 1999	March 26, 1999	September 22, 1999
State	IA	IN	GA
Statute	EPCRA	EPCRA CERCLA	EPCRA CERCLA
SEP Amount	12,000	60,000	22,167
Penalty	2,500	2,200	5,170
P2 Type(s)	Equipment/Technology Modifications	Equipment/Technology Modifications Process Modifications	Equipment/Technology Modifications

APPENDIX D, CONT'D

POLLUTION PREVENTION SUPPLEMENTAL ENVIRONMENTAL PROJECTS IN FISCAL YEAR 1999			
Order	Allied Signal, Inc.	Atlantic Richfield	Barber Dairies
Final Order Entered	September 2, 1999	July 12, 1999	September 30, 1999
State	TN	CA	AL
Statute	CAA	CAA	EPCRA CERCLA
SEP Amount	47,189	92,750	6,961
Penalty	50,238	31,990	2,320
P2 Type(s)	Process Modifications	Energy Efficiency/Conservation	Equipment/Technology Modifications

POLLUTION PREVENTION SUPPLEMENTAL ENVIRONMENTAL PROJECTS IN FISCAL YEAR 1999			
Order	Bertch Cabinet, Waterloo	Boston Edison Pilgrim Power	C & L Electric Co.
Final Order Entered	November 4, 1998	December 16, 1998	February 28, 1999
State	IA	MA	AR
Statute	RCRA	RCRA	TSCA
SEP Amount	74,000	155,247	264,049
Penalty	25,820	172,837	6,545
P2 Type(s)	Equipment/Technology Modifications	Process Modifications Improvements	Process Modifications Improvements

POLLUTION PREVENTION SUPPLEMENTAL ENVIRONMENTAL PROJECTS IN FISCAL YEAR 1999			
Order	Carmet Company	Caterpillar, Inc.	Chem-Rex
Final Order Entered	September 16, 1999	July 1, 1999	February 23, 1999
State	SC	DC	MI
Statute	EPCRA	CAA	CAA
SEP Amount	73,610	35,000,000	24,700
Penalty	26,109	25,000,000	42,000
P2 Type(s)	Equipment/Technology Modifications	Equipment/Technology Modifications Product Reformulation	Equipment/Technology Modifications

APPENDIX D, CONT'D

POLLUTION PREVENTION SUPPLEMENTAL ENVIRONMENTAL PROJECTS IN FISCAL YEAR 1999			
Order	Coastal Eagle	College Hill Poultry, Inc.	Compania Petroleum Carribbean
Final Order Entered	November 25, 1998	June 19, 1999	February 23, 1999
State	NJ	PA	PR
Statute	CAA	CWA	RCRA
SEP Amount	860,000	103,000	215,000
Penalty	300,000	17,500	80,000
P2 Type(s)	Equipment/Technology Modifications	Process Modifications	Equipment/Technology Modifications

POLLUTION PREVENTION SUPPLEMENTAL ENVIRONMENTAL PROJECTS IN FISCAL YEAR 1999			
Order	Conagra, Inc.	Cook Composites & Polymers	Cummins Engine Co.
Final Order Entered	October 1, 1998	October 15, 1998	July 1, 1999
State	ID	VA	DC
Statute	CWA	RCRA	CAA
SEP Amount	1,500,000	50,350	35,000,000
Penalty	1,000,000	109,897	25,000,000
P2 Type(s)	Equipment/Technology Modifications Process Modifications	Process Modifications Improvements	Equipment/Technology Modifications Product Reformulation

POLLUTION PREVENTION SUPPLEMENTAL ENVIRONMENTAL PROJECTS IN FISCAL YEAR 1999			
Order	Delken Company	Department of the Army	Depere Foundry
Final Order Entered	January 7, 1999	August 10, 1999	June 30, 1999
State	MA	VA	WI
Statute	EPCRA CERCLA	RCRA	CAA
SEP Amount	40,992	27,594	76,255
Penalty	8,168	6,372	6,500
P2 Type(s)	Equipment/Technology Modifications Improvements	Equipment/Technology Modifications Material Substitution	Equipment/Technology Modifications

APPENDIX D, CONT'D

POLLUTION PREVENTION SUPPLEMENTAL ENVIRONMENTAL PROJECTS IN FISCAL YEAR 1999			
Order	Detroit Diesel Co.	Discovery Energy	District of Columbia
Final Order Entered	July 1, 1999	April 22, 1999	November 13, 1998
State	DC	OK	DC
Statute	CAA	SDWA	CWA
SEP Amount	12,000,000	2,050	25,000
Penalty	12,500,000	330	5,000
P2 Type(s)	Equipment/Technology Modifications Product Reformulation	Equipment/Technology Modifications	Equipment/Technology Modifications Process Modifications

POLLUTION PREVENTION SUPPLEMENTAL ENVIRONMENTAL PROJECTS IN FISCAL YEAR 1999			
Order	E.P. Graphics	Eagle Iron Works	EG&G Wright Components
Final Order Entered	September 27, 1999	July 1, 1999	January 12, 1999
State	IN	IA	NY
Statute	CAA	TSCA	CWA
SEP Amount	108,000	10,602	76,344
Penalty	44,000	2,750	39,580
P2 Type(s)	Equipment/Technology Modifications Process Modifications Product Reformulation	Equipment/Technology Modifications	Equipment/Technology Modifications Product Reformulation

POLLUTION PREVENTION SUPPLEMENTAL ENVIRONMENTAL PROJECTS IN FISCAL YEAR 1999			
Order	Federal Bureau of Prisons	Federal Bureau of Prisons	Hall Street Realty
Final Order Entered	September 23, 1999	September 23, 1999	December 2, 1998
State	NC	NC	MO
Statute	CAA	CAA	TSCA
SEP Amount	193,109	272,333	113,400
Penalty	21,700	38,500	17,000
P2 Type(s)	Energy Efficiency/Conservation	Energy Efficiency/Conservation	Equipment/Technology Modifications

APPENDIX D, CONT'D

POLLUTION PREVENTION SUPPLEMENTAL ENVIRONMENTAL PROJECTS IN FISCAL YEAR 1999			
Order	Hartford MDC Vehicle Maintenance	Hickson Corporation	Hudson Foods, Inc.
Final Order Entered	August 25, 1999	September 17, 1999	November 17, 1998
State	CT	GA	MN
Statute	RCRA	RCRA	CERCLA
SEP Amount	5,250	787,000	31,500
Penalty	7,000	161,000	8,500
P2 Type(s)	Process Modifications Material Substitution	Process Modifications Process Recycling	Equipment/Technology Modifications

POLLUTION PREVENTION SUPPLEMENTAL ENVIRONMENTAL PROJECTS IN FISCAL YEAR 1999			
Order	Iowa Steel & Wire	Jame Fine Chemical	Johnson Controls
Final Order Entered	May 14, 1999	March 5, 1999	June 9, 1999
State	IA	NJ	IL
Statute	RCRA	EPCRA	CERCLA
SEP Amount	177,850	149,000	58,900
Penalty	15,814	15,000	2,436
P2 Type(s)	Equipment/Technology Modifications	Process Modifications	Equipment/Technology Modifications

POLLUTION PREVENTION SUPPLEMENTAL ENVIRONMENTAL PROJECTS IN FISCAL YEAR 1999			
Order	Kaiser-Francis Oil Co.	Kinze Manufacturing	Koch Fertilizer Storage
Final Order Entered	January 21, 1999	January 25, 1999	February 9, 1999
State	MI	IA	IN
Statute	SDWA	EPCRA	CERCLA
SEP Amount	2,900	181,000	130,000
Penalty	15,150	2,200	22,330
P2 Type(s)	Equipment/Technology Modifications	Equipment/Technology Modifications Process Modifications Product Reformulation	Equipment/Technology Modifications

APPENDIX D, CONT'D

POLLUTION PREVENTION SUPPLEMENTAL ENVIRONMENTAL PROJECTS IN FISCAL YEAR 1999			
Order	Lamb-Weston, Inc.	Lincoln Industrial Corp.	Lion Oil Company
Final Order Entered	October 7, 1998	April 9, 1999	September 20, 1999
State	ID	MO	AR
Statute	CAA	TSCA	CAA
SEP Amount	135,221	151,976	587,635
Penalty	160,000	26,654	90,000
P2 Type(s)	Equipment/Technology Modifications Process Modifications Process Recycling	Equipment/Technology Modifications	Process Recycle Energy Efficiency/Conservation

POLLUTION PREVENTION SUPPLEMENTAL ENVIRONMENTAL PROJECTS IN FISCAL YEAR 1999			
Order	Mack Trucks, Inc.	Massachusetts REF	McDonnell Douglas
Final Order Entered	July 1, 1999	April 2, 1999	December 8, 1998
State	DC	MA	MO
Statute	CAA	CAA	TSCA
SEP Amount	18,000,000	91,000	13,963
Penalty	13,000,000	25,000	5,500
P2 Type(s)	Equipment/Technology Modifications Product Reformulation	Process Modifications	Equipment/Technology Modifications

POLLUTION PREVENTION SUPPLEMENTAL ENVIRONMENTAL PROJECTS IN FISCAL YEAR 1999			
Order	McLaughlin Body	Modern Plating	Nature's Bounty
Final Order Entered	June 7, 1999	January 28, 1999	September 22, 1999
State	IL	OK	NY
Statute	RCRA	RCRA	EPCRA
SEP Amount	41,255	76,319	22,118
Penalty	62,489	3,933	45,000
P2 Type(s)	Process Modifications	Equipment/Technology Modifications	Materials Substitution

APPENDIX D, CONT'D

POLLUTION PREVENTION SUPPLEMENTAL ENVIRONMENTAL PROJECTS IN FISCAL YEAR 1999			
Order	Olin Corporation	Pfizer Inc.	Ponce Airlines
Final Order Entered	April 6, 1999	March 16, 1999	May 13, 1999
State	IN	CT	PR
Statute	EPCRA	RCRA EPCRA CWA	RCRA
SEP Amount	276,304	150,000	6,000
Penalty	52,604	625,000	5,000
P2 Type(s)	Equipment/Technology Modifications	Equipment/Technology Modifications Improvements	Equipment/Technology Modifications

POLLUTION PREVENTION SUPPLEMENTAL ENVIRONMENTAL PROJECTS IN FISCAL YEAR 1999			
Order	Puerto Rico Electric	Redstone Arsenal	Royal Crown Bottling Co.
Final Order Entered	March 19, 1999	December 10, 1998	February 23, 1999
State	PR	AL	IL
Statute	CAA CWA RCRA EPCRA CERCLA	SDWA	EPCRA
SEP Amount	3,500,000	807,000	24,004
Penalty	1,500,000	80,000	8,248
P2 Type(s)	Equipment/Technology Modifications Process Modifications Improvements	Equipment/Technology Modifications Improvements	Equipment/Technology Modifications

POLLUTION PREVENTION SUPPLEMENTAL ENVIRONMENTAL PROJECTS IN FISCAL YEAR 1999			
Order	State of Ind. Dep't of Corrections	Stiefel Laboratory	Stoneco, Inc.
Final Order Entered	November 5, 1998	June 7, 1999	October 8, 1998
State	IN	NY	MI
Statute	TSCA	EPCRA	CAA
SEP Amount	30,000	25,864	22,418
Penalty	3,000	8,820	3,375
P2 Type(s)	Equipment/Technology Modifications	Equipment/Technology Modifications	Process Modifications Improvements

APPENDIX D, CONT'D

POLLUTION PREVENTION SUPPLEMENTAL ENVIRONMENTAL PROJECTS IN FISCAL YEAR 1999			
Order	Stupp Corporation	Tomen Agro	U.S. Department of Army
Final Order Entered	October 14, 1998	August 9, 1999	August 18, 1999
State	LA	OH	DC
Statute	RCRA	CAA	RCRA
SEP Amount	28,487	90,200	1,000,035
Penalty	12,027	11,514	50,400
P2 Type(s)	Equipment/Technology Modifications Materials Substitution	Equipment/Technology Modifications	Equipment/Technology Modifications Process Modifications Improvements

POLLUTION PREVENTION SUPPLEMENTAL ENVIRONMENTAL PROJECTS IN FISCAL YEAR 1999			
Order	U.S. Department of Defense	U.S. Department of Energy	Utilicorp United
Final Order Entered	September 29, 1999	August 6, 1999	April 28, 1999
State	KY	NY	KS
Statute	RCRA	RCRA	CWA
SEP Amount	167,000	104,300	31,155
Penalty	125,000	17,500	2,000
P2 Type(s)	Equipment/Technology Modifications Materials Substitution	Equipment/Technology Modifications Process Modifications Materials Substitution	Equipment/Technology Modifications

POLLUTION PREVENTION SUPPLEMENTAL ENVIRONMENTAL PROJECTS IN FISCAL YEAR 1999			
Order	Vermeer Manufacturer	Vintage Petroleum	Volvo Truck Corporation
Final Order Entered	January 6, 1999	September 1, 1999	July 1, 1999
State	IA	LA	DC
Statute	EPCRA	CWA	CAA
SEP Amount	1,277,459	81,794	9,000,000
Penalty	110,442	27,734	5,000,000
P2 Type(s)	Energy Efficiency/Conservation	Equipment/Technology Modifications	Equipment/Technology Modifications Product Reformulation

APPENDIX E

DECREES, ORDERS, AND SETTLEMENTS REQUIRING FEATURES OF A COMPLIANCE-FOCUSED ENVIRONMENTAL MANAGEMENT SYSTEM

***KEY*:**

As used in this context, the use of the terms "Yes" or "No" means that the item is addressed explicitly in the consent decree or order.

As used in this context, the use of the term "presumably" means that the item is not addressed explicitly, but can be inferred to a high degree of certainty from the language of the consent decree or order.

As used in this context, the use of the term "Not Specified" means that the item is not addressed in the consent decree or order.

As used in this context, N/A is used to abbreviate the term "Not Applicable".

As used in this table, the acronym "CEM" means continuous emissions monitor, "COM" means continuous opacity monitor, "EMS" means environmental management system, "NESHAP" means National Emission Standard for Hazardous Air Pollutants, and "SEP" means Supplemental Environmental Project.

Environmental laws addressed herein include the following: Clean Air Act ("CAA"), 42 U.S.C. §7401 *et seq.* (1998); Clean Water Act ("CWA"), 33 U.S.C. §1251 *et seq.* (1998); Comprehensive Environmental Response, Compensation, and Liability Act of 1980 ("CERCLA"), 42 U.S.C. §9601 *et seq.* (1998); Emergency Response and Community Right-To-Know Act of 1986 ("EPCRA"), 42 U.S.C. §11001 *et seq.* (1998); Endangered Species Act of 1973 ("ESA"), 16 U.S.C. §1531 *et seq.* (1998); Federal Insecticide, Fungicide, and Rodenticide Act, 7 U.S.C. §136 *et seq.* (1998); Resource Conservation and Recovery Act ("RCRA"), 42 U.S.C. §6901 *et seq.* (1998); Toxic Substances Control Act ("TSCA"), 15 U.S.C. §2601 *et seq.* (1998).

APPENDIX E, CONT'D

DECREES, ORDERS, AND SETTLEMENTS REQUIRING FEATURES OF A COMPLIANCE-FOCUSED ENVIRONMENTAL MANAGEMENT SYSTEM			
I. Federal Court Decrees, Orders, and Settlements			
Decree, Order or Settlement	U.S. v. ASARCO Inc.	U.S. v. B.P. Exploration (Alaska) Inc. (Plea Agreement)	U.S. v. City of Roanoke, Virginia
Court	U.S. District Courts for the District of Montana and the District of Arizona	U.S. District Court for the District of Alaska	U.S. District Court for the Western District of Virginia, Roanoke Division
Action No.	CV-98-3-H-CCL (MT); CIV-98-0137 (AZ)	A99-0141 CR	Criminal Action 7:00CR00001-001
Date Entered	May 5, 1998 (MT); June 2, 1998 (AZ)	September 23, 1999	Plea agreement filed January 12, 2000
Violations Alleged	RCRA, CWA	CERCLA	RCRA
Specific Reference to EMS	yes	yes	yes
Provision Requiring ISO 14001 Certification	no	no	no
Provision Requiring ISO 14001 Implementation	no	no, but the EMS is based on ISO 14001 principles	no
Environmental Metrics Reporting	yes	no	no
Technology Transfer	yes	no	no
Consultant Requirements	yes	yes	yes
Third-Party Audit	yes	yes	yes
EMS Plan Approval	yes	yes	no, but an environmental compliance manual must be submitted for approval to the Probation Officer and the United States
Management Policies and Procedures: a) **Corporate Environmental Policy** b) **Site Specific Environmental Policies and Standards**	a) yes b) yes	a) yes b) yes	a) not specified b) presumably yes
Organization, Personnel and Oversight of EMS	not specified	yes	yes
Accountability and Responsibility	yes	yes	yes

APPENDIX E, CONT'D

DECREES, ORDERS, AND SETTLEMENTS REQUIRING FEATURES OF A COMPLIANCE-FOCUSED ENVIRONMENTAL MANAGEMENT SYSTEM			
I. Federal Court Decrees, Orders, and Settlements			
Environmental Requirements: a) Communication b) Interpretation c) Substantive Requirements	a) yes b) yes c) yes	a) yes b) yes c) yes	a) yes b) not specified c) yes
Assessment, Prevention and Control	yes	yes	not specified
Environment Incident and Non-compliance Investigation	yes	yes	not specified
Environmental Training, Awareness and Competence	yes[1]	yes	yes
Planning for Environmental Matters	yes	yes	not specified
Maintenance of Records and Documentation and EMS Data Protocols	yes	yes	not specified
Pollution Prevention	yes	yes	not specified
Continuing Program Evaluation	yes	yes	yes, annual compliance audits
Public Involvement/ Community Outreach	Defendant shall share the training protocol and lessons learned with other businesses and organizations through seminars	yes	not specified

APPENDIX E, CONT'D

DECREES, ORDERS, AND SETTLEMENTS REQUIRING FEATURES OF A COMPLIANCE-FOCUSED ENVIRONMENTAL MANAGEMENT SYSTEM			
I. Federal Court Decrees, Orders, and Settlements			
Decree, Order or Settlement	U.S. and Timothy R.E. Keeney, Commissioner of Environmental Protection of the State of Connecticut v. The Dexter Corporation	U.S. and State of Texas v. Encycle/ASARCO, Inc.[2]	U.S. v. FMC Corp.[3]
Court	U.S. District Court for the District Connecticut	U.S. District Court for the Southern District of Texas	U.S. District Court for the District of Idaho
Action No.	H-89-393	H-99-1136	RCRA Docket 98-0406-E-BLW
Date Entered	November 13, 1992	October 6, 1999	July 13, 1999
Violations Alleged	RCRA, CWA	RCRA, CWA	RCRA
Specific Reference to EMS	yes	yes	yes
Provision Requiring ISO 14001 Certification	no	no	no
Provision Requiring ISO 14001 Implementation	no	yes, as a practical matter	no, but the EMS is based on ISO 14001 principles
Environmental Metrics Reporting	no	no	no
Technology Transfer	no	no	no
Consultant Requirements	yes	yes	yes
Third-Party Audit	yes	yes	yes
EMS Plan Approval	yes	yes	yes, but it is not clear who approves the plan
Management Policies and Procedures: **a) Corporate Environmental Policy** **b) Site Specific Environmental Policies and Standards**	a) not specified b) not specified	a) yes b) yes	a) yes b) yes
Organization, Personnel and Oversight of EMS	not specified	not specified	yes
Accountability and Responsibility	yes	yes	yes
Environmental Requirements: **a) Communication** **b) Interpretation** **c) Substantive Requirements**	a) yes b) not specified c) yes	a) yes b) not specified c) yes	a) yes b) yes c) yes
Assessment, Prevention and Control	yes	yes	yes[4]

APPENDIX E, CONT'D

DECREES, ORDERS, AND SETTLEMENTS REQUIRING FEATURES OF A COMPLIANCE-FOCUSED ENVIRONMENTAL MANAGEMENT SYSTEM			
I. Federal Court Decrees, Orders, and Settlements			
Environment Incident and Non-compliance Investigation	yes	yes	yes
Environmental Training, Awareness and Competence	not specified	yes	yes
Planning for Environmental Matters	yes	yes	yes[5]
Maintenance of Records and Documentation and EMS Data Protocols	yes[6]	yes	yes[7]
Pollution Prevention	not specified	yes	yes
Continuing Program Evaluation	not specified	yes (audits)	yes
Public Involvement/ Community Outreach	not specified	yes	yes[8]

APPENDIX E, CONT'D

DECREES, ORDERS, AND SETTLEMENTS REQUIRING FEATURES OF A COMPLIANCE-FOCUSED ENVIRONMENTAL MANAGEMENT SYSTEM			
I. Federal Court Decrees, Orders, and Settlements			
Decree, Order or Settlement	U.S. v. HCI-Chemtech[9]	U.S. v. John Morrell & Co.[10]	U.S. v. Trident Seafoods Corporation[11]
Court	U.S. District Court for the Western District of Missouri	U.S. District Court for the District of South Dakota	U.S. District Court of Alaska
Action No.	Criminal Action	96-4011	A97-0093CV(HRH)
Date Entered	Plea agreement filed on October 30, 1997	February 11, 1998	February 24, 1998
Violations Alleged	CWA	CWA	CWA
Specific Reference to EMS	yes[12]	yes	yes
Provision Requiring ISO 14001 Certification	no	no	no
Provision Requiring ISO 14001 Implementation	no	no	no
Environmental Metrics Reporting	yes	no	no
Technology Transfer	yes?	no	no
Consultant Requirements	yes	yes	yes
Third-Party Audit	yes	yes	yes
EMS Plan Approval	yes	yes	yes
Management Policies and Procedures: a) Corporate Environmental Policy b) Site Specific Environmental Policies and Standards	a) not specified b) yes[13]	a) not specified b) yes	a) not specified b) not specified
Organization, Personnel and Oversight of EMS	yes	not specified	not specified
Accountability and Responsibility	yes	yes	not specified
Environmental Requirements: a) Communication b) Interpretation c) Substantive Requirements	a) not specified b) not specified c) yes	a) not specified b) not specified c) not specified	a) yes b) not specified c) not specified
Assessment, Prevention and Control	yes[14]	not specified	not specified

APPENDIX E, CONT'D

DECREES, ORDERS, AND SETTLEMENTS REQUIRING FEATURES OF A COMPLIANCE-FOCUSED ENVIRONMENTAL MANAGEMENT SYSTEM			
I. Federal Court Decrees, Orders, and Settlements			
Environment Incident and Non-compliance Investigation	not specified	yes	yes
Environmental Training, Awareness and Competence	yes[15]	yes	yes[16]
Planning for Environmental Matters	yes[17]	not specified	not specified
Maintenance of Records and Documentation and EMS Data Protocols	yes[18]	yes	yes
Pollution Prevention	not specified	not specified	not specified
Continuing Program Evaluation	not specified	presumably yes?	yes (audits)
Public Involvement/ Community Outreach	not specified	not specified	not specified

APPENDIX E, CONT'D

DECREES, ORDERS, AND SETTLEMENTS REQUIRING FEATURES OF A COMPLIANCE-FOCUSED ENVIRONMENTAL MANAGEMENT SYSTEM			
I. Federal Court Decrees, Orders, and Settlements			
Decree, Order or Settlement	U.S. v. Trustees of Boston University	U.S. and Timothy R.E. Keeney, Commissioner of the Department of Environmental Protection of the State of Connecticut v. United Technologies Corporation[19]	U.S. v. USX Corporation[20]
Court	U.S. District Court for the District of Massachusetts	U.S. District Court of Connecticut	U.S. District Court for the Northern District of Indiana
Action No.	97-12261-PBS	H-90-715 (JAC)	2:98CV-465JM
Date Entered	December 9, 1997	August 23, 1993	January 26, 1999
Violations Alleged	CWA, RCRA	RCRA, CWA	CWA, RCRA
Specific Reference to EMS	yes	yes	yes
Provision Requiring ISO 14001 Certification	no	no	no
Provision Requiring ISO 14001 Implementation	no	no	no
Environmental Metrics Reporting	no	no	yes
Technology Transfer	no	no	no
Consultant Requirements	yes	yes	no
Third-Party Audit	yes	yes	no
EMS Plan Approval	yes	yes	no
Management Policies and Procedures: a) Corporate Environmental Policy b) Site Specific Environmental Policies and Standards	a) not specified b) not specified	a) not specified b) not specified	a) not specified b) not specified
Organization, Personnel and Oversight of EMS	yes	yes	yes
Accountability and Responsibility	yes	yes	yes
Environmental Requirements: a) Communication b) Interpretation c) Substantive Requirements	a) not specified b) not specified c) not specified	a) not specified b) not specified c) yes	a) yes b) not specified c) yes

APPENDIX E, CONT'D

DECREES, ORDERS, AND SETTLEMENTS REQUIRING FEATURES OF A COMPLIANCE-FOCUSED ENVIRONMENTAL MANAGEMENT SYSTEM			
I. Federal Court Decrees, Orders, and Settlements			
Assessment, Prevention and Control	yes	yes	yes
Environment Incident and Non-compliance Investigation	yes	yes	yes
Environmental Training, Awareness and Competence	yes	yes	yes
Planning for Environmental Matters	not specified	not specified	yes
Maintenance of Records and Documentation and EMS Data Protocols	not specified	not specified	yes no EMS data protocols
Pollution Prevention	not specified	yes	not specified
Continuing Program Evaluation	not specified	not specified	yes
Public Involvement/ Community Outreach	not specified	yes	yes

APPENDIX E, CONT'D

DECREES, ORDERS, AND SETTLEMENTS REQUIRING FEATURES OF A COMPLIANCE-FOCUSED ENVIRONMENTAL MANAGEMENT SYSTEM			
I. Federal Court Decrees, Orders, and Settlements			
Decree, Order or Settlement	U.S. v. Weirton Steel Corporation[21]	Natural Resources Defense Council, Inc., and U.S. v. Wheeling-Pittsburgh Steel Corporation (Allenport facility)	U.S. v. Wheeling-Pittsburgh Steel Corporation (Steubenville, Mingo Junction and Yorkville facilities)
Court	U.S. District Court for the Northern District of West Virginia	U.S. District Court for the Western District of Pennsylvania	U.S. District Court for the Southern District of Ohio, Eastern Division
Action No.	5:96-CV-171	89-237 S	C288-598
Date Entered	October 31, 1996	November 4, 1992	July 16, 1991
Violations Alleged	CAA, CWA, RCRA	CWA	CWA
Specific Reference to EMS	yes	yes	yes
Provision Requiring ISO 14001 Certification	no	no	no
Provision Requiring ISO 14001 Implementation	no	no	no
Environmental Metrics Reporting	no	no	no
Technology Transfer	no	no	no
Consultant Requirements	yes	yes	no
Third-Party Audit	yes	yes	yes
EMS Plan Approval	yes	yes	yes
Management Policies and Procedures: a) Corporate Environmental Policy b) Site Specific Environmental Policies and Standards	a) yes b) yes	a) yes b) not specified	a) yes b) not specified
Organization, Personnel and Oversight of EMS	yes	not specified	not specified
Accountability and Responsibility	yes	yes	yes
Environmental Requirements: a) Communication b) Interpretation c) Substantive Requirements	a) yes b) yes c) yes	a) yes b) not specified c) not specified	a) yes b) not specified c) not specified

APPENDIX E, CONT'D

DECREES, ORDERS, AND SETTLEMENTS REQUIRING FEATURES OF A COMPLIANCE-FOCUSED ENVIRONMENTAL MANAGEMENT SYSTEM			
I. Federal Court Decrees, Orders, and Settlements			
Assessment, Prevention and Control	yes	not specified	not specified
Environment Incident and Non-compliance Investigation	yes	not specified	not specified
Environmental Training, Awareness and Competence	yes	not specified	not specified
Planning for Environmental Matters	yes	not specified	not specified
Maintenance of Records and Documentation and EMS Data Protocols	yes no EMS data protocol	not specified	not specified
Pollution Prevention	yes	not specified	not specified
Continuing Program Evaluation	yes	not specified	not specified
Public Involvement/ Community Outreach	not specified	not specified	not specified

APPENDIX E, CONT'D

DECREES, ORDERS, AND SETTLEMENTS REQUIRING FEATURES OF A COMPLIANCE-FOCUSED ENVIRONMENTAL MANAGEMENT SYSTEM			
II. Federal Administrative Orders and Agreements			
Order or Agreement	In the Matter of Chem-Security Systems, Inc.	MOA Between United States of America and the Department of Energy (Brookhaven National Laboratory)[22]	
Court	N/A	N/A	
Civil Action No.	1085-06-08-3008P	(on file with authors)	
Date Entered	July 18, 1985	March 23, 1998	
Violations Alleged	RCRA	RCRA	
Specific Reference to EMS	no, but to a compliance management system	yes	
Provision Requiring ISO 14001 Certification	no	no	
Provision Requiring ISO 14001 Implementation	no	no, but the EMS is based on ISO 14001 principles	
Environmental Metrics Reporting	no	no	
Technology Transfer	no	no	
Consultant Requirements	no	no	
Third-Party Audit	no	yes	
EMS Plan Approval	no	no	
Management Policies and Procedures: **a) Corporate Environmental Policy** **b) Site Specific Environmental Policies and Standards**	a) not specified b) not specified	a) yes b) not specified	
Organization, Personnel and Oversight of EMS	no	yes	
Accountability and Responsibility	yes	yes	
Environmental Requirements: **a) Communication** **b) Interpretation** **c) Substantive Requirements**	a) yes b) not specified c) yes	a) yes b) not specified c) yes	

APPENDIX E, CONT'D

DECREES, ORDERS, AND SETTLEMENTS REQUIRING FEATURES OF A COMPLIANCE-FOCUSED ENVIRONMENTAL MANAGEMENT SYSTEM			
II. Federal Administrative Orders and Agreements			
Assessment, Prevention and Control	yes	yes	
Environment Incident and Non-compliance Investigation	yes	yes	
Environmental Training, Awareness and Competence	not specified	yes	
Planning for Environmental Matters	not specified	yes	
Maintenance of Records and Documentation and EMS Data Protocols	presumably yes	presumably yes	
Pollution Prevention	not specified	yes	
Continuing Program Evaluation	yes[23]	yes	
Public Involvement/ Community Outreach	not specified	yes	

APPENDIX E, CONT'D

DECREES, ORDERS, AND SETTLEMENTS REQUIRING FEATURES OF A COMPLIANCE-FOCUSED ENVIRONMENTAL MANAGEMENT SYSTEM			
III. State Court Decrees and Orders			
Decree or Order	State of Arizona v. Arizona Water Company[24]	State of California v. San Diego Farah Partners and Behram Baxter	The Department of Natural Resources and Environmental Control v. Nanticoke Homes, Inc.[25]
Court	Superior Court of Arizona, County of Coconino	Superior Court of California, County of San Diego	Superior Court for the State of Delaware in and for New Castle County
Civil Action No.	CV98-0038	(on file with authors)	98M-02-072 JOH
Date Entered	January 23, 1998	June 1999	February 12, 1998
Violations Alleged	Violations of rules of the Arizona Department of Environmental Quality governing the discharge of pollutants into the waters of the State of Arizona.	Violation of local Health and Safety Code	Delaware Code Chapters 60 (air quality), 63 (hazardous waste), and 74 (underground storage tanks) Delaware Air Pollution Control, Hazardous Waste, and UST Regulations
Specific Reference to EMS	yes	yes	yes
Provision Requiring ISO 14001 Certification	no	yes	no
Provision Requiring ISO 14001 Implementation	no	no	no
Environmental Metrics Reporting	no	no	no
Technology Transfer	no	yes	no
Consultant Requirements	no	no	yes
Third-Party Audit	no	yes	no
EMS Plan Approval	no	yes	no
Management Policies and Procedures: a) Corporate Environmental Policy b) Site Specific Environmental Policies and Standards	a) yes[26] b) not specified	a) not specified b) not specified	a) presumably yes b) not specified
Organization, Personnel and Oversight of EMS	yes[27]	not specified	presumably yes
Accountability and Responsibility	yes[28]	not specified	presumably yes

APPENDIX E, CONT'D

DECREES, ORDERS, AND SETTLEMENTS REQUIRING FEATURES OF A COMPLIANCE-FOCUSED ENVIRONMENTAL MANAGEMENT SYSTEM			
III. State Court Decrees and Orders			
Environmental Requirements: a) Communication b) Interpretation c) Substantive Requirements	a) yes[29] b) not specified c) yes	a) not specified b) not specified c) not specified	a) not specified b) not specified c) yes
Assessment, Prevention and Control	yes[30]	not specified	presumably yes
Environment Incident and Non-compliance Investigation	yes	not specified	presumably yes
Environmental Training, Awareness and Competence	yes[31]	yes	yes; in addition, $25,000 shall be paid to provide information to the Management Consultant to train and educate hourly workers
Planning for Environmental Matters	not specified	not specified	not specified
Maintenance of Records and Documentation and EMS Data Protocols	not specified	not specified	yes, Defendant is to preserve all records for at least five years after entry of the Consent Order
Pollution Prevention	not specified	not specified	yes?
Continuing Program Evaluation	yes, unannounced internal inspections	yes (audits)	not specified
Public Involvement/ Community Outreach	not specified	yes	Pay $40,000 to support and fund the Department of Natural Resources and Environmental Control Clean Air Act and Public Outreach Program

APPENDIX E, CONT'D

DECREES, ORDERS, AND SETTLEMENTS REQUIRING FEATURES OF A COMPLIANCE-FOCUSED ENVIRONMENTAL MANAGEMENT SYSTEM			
III. State Court Decrees and Orders			
Decree or Order	Commissioner of the Department of Environment v. Reliance Electric-Highland[32]		
Court	Indiana Department of Environmental Management		
Civil Action No.	Cause No. H-13028		
Date Entered	January 16, 1998		
Violations Alleged	Violations of 40 C.F.R. §262.11, 40 C.F.R. Part 264 and 270, 40 C.F.R. §268.50(a)(1), 329 IAC §3.1-15-4. 329 IAC §3.1-15-8, 40 C.F.R. §265.31 governing the operation and storage of solid waste		
Specific Reference to EMS	yes[33]		
Provision Requiring ISO 14001 Certification	yes		
Provision Requiring ISO 14001 Implementation	yes		
Environmental Metrics Reporting	yes		
Technology Transfer	yes		
Consultant Requirements	no		
Third-Party Audit	no		
EMS Plan Approval	no		
Management Policies and Procedures: **a) Corporate Environmental Policy** **b) Site Specific Environmental Policies and Standards**	a) presumeably yes b) Defendant shall develop an environmental policy for its facility		
Organization, Personnel and Oversight of EMS	yes[34]		
Accountability and Responsibility	yes[35]		

APPENDIX E, CONT'D

DECREES, ORDERS, AND SETTLEMENTS REQUIRING FEATURES OF A COMPLIANCE-FOCUSED ENVIRONMENTAL MANAGEMENT SYSTEM			
III. State Court Decrees and Orders			
Environmental Requirements: a) Communication b) Interpretation c) Substantive Requirements	a) yes[36] b) not specified c) yes		
Assessment, Prevention and Control	yes[37]		
Environment Incident and Non-compliance Investigation	yes		
Environmental Training, Awareness and Competence	Train employees to increase awareness and competency		
Planning for Environmental Matters	yes		
Maintenance of Records and Documentation and EMS Data Protocols	yes		
Pollution Prevention	not specified		
Continuing Program Evaluation	yes		
Public Involvement/ Community Outreach	yes[38]		

APPENDIX E, CONT'D

DECREES, ORDERS, AND SETTLEMENTS REQUIRING FEATURES OF A COMPLIANCE-FOCUSED ENVIRONMENTAL MANAGEMENT SYSTEM			
IV. State Administrative Orders			
Order	Donsco, Inc.[39]	In the Matter of General Motors Corporation Wilmington Assembly Plant[40]	Hussey Copper Ltd.[41]
Agency	Commonwealth of Pennsylvania Department of Environmental Protection	State of Delaware Department of Natural Resources and Environmental Control	Commonwealth of Pennsylvania Department of Environmental Protection
Docket No.	(not present on order)	(not present on order)	Docket No. 5-97015
Date Entered	April 15, 1998	June 16, 1998	November 17, 1997
Violations Alleged	Pennsylvania Air Pollution Control Act, Pennsylvania Clean Streams Law, Pennsylvania Solid Waste Management Act, Pennsylvania Administrative Code, Pennsylvania Storage Tank and Spill Prevention Act	Delaware Code Chapter 60 Delaware Air Pollution Control Regulations	Pennsylvania Solid Waste Management Act, Pennsylvania Clean Streams Law, Pennsylvania Administrative Code
Specific Reference to EMS	yes	yes[42]	yes
Provision Requiring ISO 14001 Certification	no	yes	no
Provision Requiring ISO 14001 Implementation	no	no	no
Environmental Metrics Reporting	no	no	no
Technology Transfer	no	no	no
Consultant Requirements	yes	no	yes
Third-Party Audits	no	no	no
EMS Plan Approval	yes	no	yes
Management Policies and Procedures: a) Corporate Environmental Policy b) Site Specific Environmental Policies and Standards	a) yes b) yes[43]	a) not specified b) yes[44]	a) yes[45] b) yes

APPENDIX E, CONT'D

DECREES, ORDERS, AND SETTLEMENTS REQUIRING FEATURES OF A COMPLIANCE-FOCUSED ENVIRONMENTAL MANAGEMENT SYSTEM			
IV. State Administrative Orders			
Organization, Personnel and Oversight of EMS	presumably yes	yes[46]	presumably yes[47]
Accountability and Responsibility	yes[48]	yes	yes[49]
Environmental Requirements: a) Communication b) Interpretation c) Substantive Requirements	a) not specified b) not specified c) yes	a) yes b) not specified c) yes	a) not specified b) not specified c) yes
Assessment, Prevention and Control	Assessment of manufacturing process and operations and pollution prevention opportunities for all sources of waste generation	presumably yes	Assessment of manufacturing process and operations and pollution prevention opportunities for all sources of waste generation
Environment Incident and Non-compliance Investigation	yes[50]	not specified	not specified
Environmental Training, Awareness and Competence	yes	not specified	Establishment of an Environmental Action Program ("EAP") which will include employee training
Planning for Environmental Matters	yes	not specified	not specified
Maintenance of Records and Documentation and EMS Data Protocols	yes	not specified	not specified
Pollution Prevention	yes	yes	yes[51]
Continuing Program Evaluation	not specified	not specified	not specified
Public Involvement/ Community Outreach	yes[52]	yes[53]	yes[54]

APPENDIX E, CONT'D

DECREES, ORDERS, AND SETTLEMENTS REQUIRING FEATURES OF A COMPLIANCE-FOCUSED ENVIRONMENTAL MANAGEMENT SYSTEM			
IV. State Administrative Orders			
Order	Keystone Cement Company[55]		
Agency	Commonwealth of Pennsylvania Department of Environmental Protection		
Docket No.	(not present on order)		
Date Entered	July 10, 1998		
Violations Alleged	Pennsylvania Air Pollution Control Act, Pennsylvania Clean Streams Law, Pennsylvania Solid Waste Management Act, Pennsylvania Administrative Code, Pennsylvania Environmental Quality Board Regulations		
Specific Reference to EMS	yes		
Provision Requiring ISO 14001 Certification	no		
Provision Requiring ISO 14001 Implementation	no		
Environmental Metrics Reporting	no		
Technology Transfer	no		
Consultant Requirements	yes		
Third-Party Audit	no		
EMS Plan Approval	yes		
Management Policies and Procedures: **a) Corporate Environmental Policy** **b) Site Specific Environmental Policies and Standards**	a) yes b) not specified		
Organization, Personnel and Oversight of EMS	yes		

APPENDIX E, CONT'D

DECREES, ORDERS, AND SETTLEMENTS REQUIRING FEATURES OF A COMPLIANCE-FOCUSED ENVIRONMENTAL MANAGEMENT SYSTEM			
IV. State Administrative Orders			
Accountability and Responsibility	yes		
Environmental Requirements: **a) Communication** **b) Interpretation** **c) Substantive Requirements**	a) yes, a process for establishing internal and external communication on environmental management issues b) not specified c) yes		
Assessment, Prevention and Control	yes		
Environment Incident and Non-compliance Investigation	yes[56]		
Environmental Training, Awareness and Competence	yes, training program to ensure that all employees are capable of performing their environmental responsibilities		
Planning for Environmental Matters	yes, as to how to respond to potential emergencies		
Maintenance of Records and Documentation and EMS Data Protocols	yes, maintenance of EMS documentation that describes Keystone's EMS and procedures to update documentation		
Pollution Prevention	not specified		
Continuing Program Evaluation	yes		
Public Involvement/ Community Outreach	yes, communicating environmental management issues to neighbors, community groups, and local officials		

APPENDIX E, CONT'D

DECREES, ORDERS, AND SETTLEMENTS REQUIRING FEATURES OF A COMPLIANCE-FOCUSED ENVIRONMENTAL MANAGEMENT SYSTEM			
V. Canadian Court Orders			
Order	Her Majesty the Queen v. Coretec	Her Majesty the Queen v. Prospec Chemicals	
Court	In the Ontario Court (Provincial Division) Toronto Region	The Provincial Court of Alberta, Judicial District of Fort Saskatchewan	
Civil Action No.	(not present on decree)	51141166P10101/P10102	
Date Entered	August 18, 1998	January 25, 1996	
Violations Alleged	Canadian Environmental Protection Act	Environmental Protection and Enhancement Act	
Specific Reference to EMS	yes[57]	yes	
Provision Requiring ISO 14001 Certification	yes	yes	
Provision Requiring ISO 14001 Implementation	no	no	
Environmental Metrics Reporting	no	no	
Technology Transfer	no	no	
Consultant Requirements	no	no	
Third-Party Audit	no	no	
EMS Plan Approval	no	no	
Management Policies and Procedures: a) Corporate Environmental Policy b) Site Specific Environmental Policies and Standards	a) Defendant shall establish an environmental policy b) not specified	a) not specified b) not specified	
Organization, Personnel and Oversight of EMS	not specified	not specified	
Accountability and Responsibility	yes	not specified	

APPENDIX E, CONT'D

DECREES, ORDERS, AND SETTLEMENTS REQUIRING FEATURES OF A COMPLIANCE-FOCUSED ENVIRONMENTAL MANAGEMENT SYSTEM			
V. Canadian Court Orders			
Environmental Requirements: a) Communication b) Interpretation c) Substantive Requirements	a) not specified b) not specified c) yes	a) not specified b) not specified c) not specified	
Assessment, Prevention and Control	yes	not specified	
Environment Incident and Non-compliance Investigation	not specified	not specified	
Environmental Training, Awareness and Competence	yes (related to training neighboring facilities)	not specified	
Planning for Environmental Matters	not specified	not specified	
Maintenance of Records and Documentation and EMS Data Protocols	not specified	not specified	
Pollution Prevention	yes (related to providing programs for neighboring facilities)	yes	
Continuing Program Evaluation	not specified	not specified	
Public Involvement/ Community Outreach	yes[58]	not specified	

NOTES TO APPENDIX E

1. Consent decree requires the defendant to develop and implement an Employee Training Program.

2. Chapter 361 of the TEX. HEALTH & SAFETY CODE, TEXAS HEALTH AND SAFETY CODE ANN. §§361.001 *et seq.*, and Chapter 7 of the Texas Water Code, TEX. WATER CODE ANN. §§7.001 *et seq.*, and the rules and regulations promulgated pursuant to those statutes.

3. In October 1998, FMC Corporation negotiated an agreement with EPA to resolve alleged violations of RCRA. United States v. FMC Corp., Civ. Action No. 98-0406-E-BLW (D. Idaho, Oct. 16, 1998) (Consent Decree). Under the EMS requirements, the company agreed to spend approximately $170 million, including the largest civil penalty ever assessed under RCRA ($11,864,800), to settle alleged hazardous waste violations. Under the consent decree, FMC is to comply with the provisions of a separate Compliance Schedule, which includes an obligation to develop an EMS for the company's Pocatello facility that is "consistent with state-of-the-art principles of environmental management." The elements of the EMS are to be derived from company-specific principles and requirements, as well as ISO 14001. In addition, the consent decree places "enhanced emphasis on compliance assurance, pollution prevention and community outreach," incorporating the 12 key elements of the NEIC's compliance-focused EMS. The consent decree calls for the appointment of an EMS Task Force composed of plant environmental, engineering and operating personnel, in-house corporate legal and environmental staff, and independent third-party consultants with EMS design expertise to evaluate the current environmental compliance program and to develop an overall EMS that will ensure compliance with environmental and operational requirements. The EMS is to be implemented within 24 months of the date of approval of the written EMS Manual. After 18 months have passed, the EMS Task Force is to arrange and fund a independent third-party EMS audit.

4. FMC agreed to submit a Quality Assurance Project Plan ("QAPP") that complies with EPA's "Interim Guidelines and Specifications for Preparing Quality Assurance Project Plans." (QAM-005/80), "Data Quality Objective Guidance" (EPA/540/G87/003). Further, the company must ensure that its laboratories analyze all samples submitted by EPA for quality in accordance with the QAPP and allow EPA permitted access.

5. FMC committed to take all appropriate action to prevent, abate or minimize release of hazardous substance, pollutant, contaminant or endangerment which threatens a release or presents an imminent and substantial endangerment to the public health or welfare or the environment in accordance with all applicable provisions of the Health and Safety Plan and the Contingency Plan submitted as part of the final Remedial Design.

6. Consent decree requires Dexter to retain at least one legible copy of all records, documents, reports or plans required by its NPDES permit or which relate to its performance under its CWA compliance program, as noted in the decree, and any underlying research or data, including computer tapes, which Dexter relied on in preparing such records, documents, reports or plans for a period of five years from the date of such a record, document, report, plan or underlying research or data, or until termination of the decree, whichever is later. In addition to any state or federal requirements relating to record retention, Dexter is obligated to retain at least one copy of all records, documents, reports, or plans which relate to its performance under its RCRA compliance program, as noted in the decree, and any underlying research or data, including computer tapes, which Dexter relied on in preparing such records, documents, reports or plans, for a period of 10 years from the date of such record, document, report, plan or underlying research and data, or until termination of the decree, whichever is later.

7. FMC must preserve records for 10 years after EPA certification of completion of work.

8. FMC must cooperate with EPA by providing information to the public regarding the work to be performed in accordance with the decree, and participate in the preparation of such information and in public meetings.

9. In 1997, HCI Chemtech, when pleading guilty to criminal felony violations of the Clean Water Act, agreed not only to pay a $175,000 fine and make a $21,000 restitution payment to the Missouri Department of Natural Resources ("MDNR"), but also committed as a special condition of a three-year probationary period to "adopt reasonable Corporate Compliance and Monitoring programs" in conjunction with the MDNR and EPA. United States v. HCI Chemtech Distribution, Inc., No. 96-00156-01-CR-W-2 (W.D. Mo. Apr. 10, 1997) (Plea Agreement). The elements of the Corporate Compliance and Monitoring programs were based on ASTM Standard E 1527-97, the ASTM Standard for Phase II Environmental Site Assessments and Environmental Management Systems. Part I of a Compliance Agreement between the company and EPA calls for the development of an Environmental Investigation Work Plan, an audit of all of the company's facilities in Missouri, a summary of all environmental investigations pertaining to the three facilities covered by the agreement, a list of all hazardous substances stored or handled on site, and soil sampling and air and groundwater monitoring. Part II of the Compliance Agreement calls for the company to establish an EMS at its Missouri facilities. The EMS was based on the "Responsible Distribution System" developed by the National Association of Chemical Distributors. It is similar to ISO 14001, but is designed for chemical distribution firms. If, after a year, the MDNR and EPA were not satisfied with the Responsible Distribution System, the company could be required to implement ISO 14001. To implement the EMS, the company must retain an independent third-party auditor to conduct an Initial EMS Review and Evaluation. Based on the results of the auditor's review, HCI must prepare a written EMS and complete an EMS Manual that meets or exceeds the provisions of ISO 14001. The Manual must be submitted to a third-party auditor for review, comment, and approval, then to EPA and the MDNR for their approval. The company must also commit sufficient financial resources to the EMS and submit monthly status reports to the MDNR, EPA, and the U.S. Office of Probation and Parole. Finally, the agreement calls for a third-party auditor to perform an EMS audit after one year, and report in writing to HCI, EPA, MDNR, and the company's probation officer. U.S. EPA Regional Counsel Martha Steincamp characterized the HCI-Chemtech matter as significant in that it "pioneers the use of criminal prosecutions to bring companies into the use of Environmental Management Systems."

10. In January 1998, John Morrell and Company negotiated a consent decree with EPA to resolve alleged violations of the Clean Water Act by the company's facility in Sioux Falls, South Dakota. United States v. John Morrell & Co., Civil Action No. 96-4011 (D.S.D. Jan. 13, 1998) (Second Partial Consent Decree). In addition to payment of a $250,000 civil penalty, Morrell is obligated under the terms of its agreement with EPA to conduct an environmental management audit ("EMA") of its Sioux Falls facility. This audit—to "assess Facility processes, internal controls (management and engineering), business unit organization at the Facility, and responsibilities of each business unit at the Facility with respect to applicable Environmental Laws"—is to be developed and performed by an independent third party approved by the Agency. The Agency is to review and approve an EMA Work Plan setting forth procedures the auditors would follow to address management compliance with environmental laws, so that the auditors and Morrell will be able to: (i) evaluate management practices governing activities undertaken at the facility that are regulated under environmental laws; (ii) identify whether management has sufficient independent authority to ensure current and future compliance, and is reliable and accountable at the appropriate corporate level; (iii) evaluate the effectiveness of systems, plans and procedures used at the facility or elsewhere to ensure compliance; (iv) recommend any necessary facility activities with respect to compliance; (v) recommend any necessary changes to systems or propose new systems if necessary, for responding to changes in environmental laws; and (vi) identify whether there exists an ongoing practice by which the company's system for achieving and maintaining compliance with environmental laws at the facility is evaluated.

Under the EMA Work Plan, the auditors are to: (i) identify all locations where activities regulated by environmental laws take place at the facil-

NOTES TO APPENDIX E (CONT'D)

ity; (ii) determine whether there are existing management and control systems in place to establish compliance; (iii) where no system currently exists, identify all factors necessary to achieve and maintain compliance; (iv) review organizational structure to determine appropriate locations for existing or new systems to manage compliance; and (v) specify a schedule for conducting the audit, submitting an EMA Audit report, and completing all tasks set forth in the work plan. The audit firm is to be granted access to all "non-privileged" records, documents, and information. In addition, the audit firm is to prepare an EMA Report providing a complete description of the company's environmental management programs, report the findings of the analysis of the EMA Report, and recommend initiatives for remedying problems identified, to EPA. Within 90 days after receipt of the EMA Report, the company is to submit an EMA Implementation Plan and Schedule ("EMAIP&S") for agency approval, indicating which recommendations it will follow (and within what time period), and the basis for any decision not to implement any recommendation. The EMAIP&S is not enforceable under the consent decree.

11. A consent decree entered into by Trident Seafoods Corporation in the U.S. District Court in Alaska in 1998 calls for the company to pay the U.S. Government a civil penalty of $418,150, reduce concentrations of the biochemical oxygen demand at selected plants, and undertake improvements in environmental compliance management to resolve alleged violations of the Clean Water Act. United States v. Trident Seafoods Corp., No. A97-0093CV (HRH) (D. Alaska Feb. 24, 1998) (Consent Decree). Specifically, the company agreed to retain an independent consultant to conduct biennial environmental audits to assess not only compliance, but also to evaluate the effectiveness of Trident's management system and practices for maintaining compliance with environmental laws. Trident agreed that within 60 days of submission of final audit reports to EPA, it would submit to the Agency for its review an environmental management systems improvement plan containing the company's proposal for implementing the auditor's recommendations.

12. Based on an initial review and evaluation of current operations, Chemtech committed to prepare a written comprehensive EMS in compliance with the company's Responsible Distribution System program for the facilities covered under the agreement.

13. The company must outline a detailed description of each facility-specific strategy in the work plan.

14. If after one year from the effective date of the compliance agreement, EPA and the MDNR determine that Chemtech is not making a substantial effort to comply with the agreement, EPA and the MDNR may seek a court order for remote site monitoring.

15. Chemtech's Health and Safety Plan/Contingency Plan must address all training required for employees, visitors; and contractors.

16. The Trident EMS Manual must identify specific education and training programs for plant personnel, as well as document any training provided.

17. *See supra* note 10.

18. Chemtech agreed to maintain an accurate inventory of all products and wastes.

19. In July 1993, United Technologies Corp. ("UTC"), EPA, and the Connecticut Department of Environmental Protection entered into a consent decree in connection with UTC's alleged violations of the Resource Conservation and Recovery Act, the Clean Water Act, and the Connecticut Water Pollution Control Act. United States v. United Tech. Corp., Civil Action No. H-90-715 (JAC) (D. Conn. July 30, 1993) (Consent Decree). The consent decree calls for UTC to pay a civil penalty of $4,251,910 to the Federal Government, and $1,050,000 to the State of Connecticut. Among other actions, the decree also calls upon the firm to provide compliance training for local small business, and comply with the terms and conditions of an attached Audit Agreement. Under the Audit Agreement, UTC committed to: (i) perform a Management Systems Analysis; (ii) submit monthly progress reports; and (iii) prepare and submit a Management Systems Report, describing the efficacy of the company's environmental management programs, analyzing proposed changes to such programs, and recommending additional ways to improve the company's EMSs, policies, and practices. A Management Systems Report would be prepared by an independent outside consultant (selected by UTC but approved by EPA), with only the consultant's recommendations going to EPA. UTC would then submit a Management Systems Improvement Plan describing follow-through on consultant's recommendations. The consultant would also perform one follow-up visit and submit a report to EPA verifying that UTC has effected changes proposed in the Management Systems Improvement Plan. UTC must also engage an independent audit firm to conduct compliance audits at selected facilities, using prescribed Protocol, and to report on specified areas.

20. In August 1998, USX Corp. and EPA entered into an agreement governing the environmental management system of its facility in Gary, Indiana. United States v. USX Corp., Civ. Action No. 2:98CV-465-JM (N.D. Ind. Aug. 6, 1998) (Consent Decree). The government had alleged violations of the Clean Water Act and the Emergency Planning and Community Right-To-Know Act, and USX had performed an analysis of its EMS for the facility at issue, and had decided upon the design of a comprehensive EMS. As part of that evaluation, and in consultation with EPA, USX prepared an EMS Manual for implementation at the facility. Under the terms of a consent decree, USX committed to use its best efforts to implement the EMS Manual pursuant to a set schedule. The company also agreed to submit annual reports to the Agency concerning its progress toward implementation, including analysis of: (i) USX's compliance with the consent decree, its NPDES permit, and §311 of the Clean Water Act; and (ii) how implementation of the EMS impacted USX's compliance record during the period under review.

21. Weirton Steel concluded an EMS agreement with EPA and the West Virginia Division of Environmental Protection ("WVDEP") in October 1996. United States v. Weirton Steel Corp., Civ. Action No. 5:96-CV-171 (N.D. W. Va. Oct. 31, 1996). The consent decree resolved allegations of violations of the Clean Air Act, the Resource Conservation and Recovery Act, the Clean Water Act, and the West Virginia Water Pollution Control Act. The company was obligated to select a qualified auditor to conduct an Initial EMS Review and Evaluation of the current EMS relating to operations at its Weirton, West Virginia ,. Based on the Initial EMS Review and Evaluation results, the company must prepare a written "Comprehensive EMS" for the facility addressing the key elements of the NEIC policy, including: (i) management policies and procedures; (ii) organization, personnel, and oversight of the EMS; (iii) accountability and responsibility; (iv) environmental requirements; (v) assessment, prevention, and control; (vi) environmental incident and non-compliance investigations; (vii) environmental education and training; (viii) planning for environmental matters; (ix) maintenance if records and documentation; (x) pollution prevention; and (xi) continuing program evaluation. Defendant would then be given 270 days from date of decree to complete an Environmental Management Systems Manual containing the Comprehensive EMS, an EMS implementation schedule and rationale identifying the proposed changes to the current EMS and explaining what the changes are intended to accomplish. EPA and WVDEP would receive an opportunity to comment on Manual. Weirton would then commence implementation of the EMS in accordance with schedule set forth in EMS Manual upon receipt of EPA's and WVDEP's' comments.

Weirton was required to report to EPA and WVDEP concerning progress of EMS implementation on a quarterly basis. Within 20 months of the decree, the company was to have selected a qualified, independent, third-party consultant to perform a Comprehensive EMS Review and Evaluation, consisting of a review and evaluation of the implementation of the systems, policies, and procedures described in the EMS Manual. The auditor would then complete the Review and Evaluation and submit a Report to Weirton, EPA, and WVDEP, containing both the results of the review and evaluation and the audi-

NOTES TO APPENDIX E (CONT'D)

tor's recommendations for improvements to the Comprehensive EMS. Weirton then was to have a limited time to respond in writing, indicating which recommendations it would, or would not, follow.

22. The Brookhaven National Laboratory ("BNL") and EPA negotiated an agreement in March of 1998, following a New York Attorney General investigation of the "high-profile Superfund site on which nuclear reactors are located and low-level radioactive waste has been released." Memorandum of Agreement By and Between the United States Environmental Protection Agency and the United States Department of Energy dated March 23, 1998 (on file with authors). The investigation uncovered structural problems with the High Flux Beam Reactor that have caused leaks of radioactive cooling water and a need for a change in practices and attitudes regarding environmental protection and public health. A request for proposals from operating contractor candidates requires demonstration of substantial and measurable progress toward environmental, safety and health managerial excellence, such as ISO 14001. Reactor start-up cannot occur until environmental management improvements are made. The Attorney General report calls for BNL to adopt "strict, cutting-edge environmental management practices that will be subject to external review and audit." The DOE/EPA Memorandum of Agreement includes a five-year Environmental Management Systems Audit Agreement setting forth goals, objectives, provisions for EMS audits and independent third-party EMS audits in the third year. BNL's integrated management system incorporates ISO 14001 and DOE safety management components and places an "enhanced emphasis on compliance assurance, pollution prevention, and community outreach." In the words of the Attorney General, "it takes no great leap to conclude that if ISO 14000 policies and practices had been firmly in place, many of the instances of environmental mismanagement . . . could have been prevented or at least minimized."

23. The company's EMS Manual must provide for at least an annual evaluation of the EMS.

24. When the Arizona Water Company ("AWC") was charged with violating state water and littering statutes it worked with the Arizona Department of Environmental Quality ("DEQ") to develop a solution. State of Arizona v. Arizona Water Co., No. CV98-0038 (Ariz. Super. Ct. Jan. 23, 1998) (Consent Judgment). In January 1998, under the terms of a Stipulated Plea Agreement and related Consent Decree, AWC agreed to agree to a settlement under which it would implement an Environmental Compliance Management System incorporating seven main elements. First, the company had to adopt a corporate policy statement regarding compliance with environmental regulations. Second, a series of company-wide, periodic environmental audits by qualified employees would be conducted. Third, the firm would establish company-wide policies governing the discipline of employees who fail to comply with environmental regulations, as well as policies governing the rewarding of employees for actions resulting in environmental improvements. Fourth, the plea agreement called for the appointment of an environmental compliance officer. Additionally, there would be a system for the establishment of a channel of communication between the compliance officer and employees, and a policy to assure employees that concerns expressed to management regarding environmental compliance would be addressed. Also, a training program would be established, encompassing environmental compliance awareness and job-specific environmental compliance. Finally, operations would be subject to periodic, unannounced environmental inspections by the compliance officer.

25. In February 1998, the Delaware Department of Natural Resources and Environmental Control ("DNREC") came to an agreement with Nanticoke Homes, Inc. regarding alleged violations of the Delaware Air Pollution Control, Hazardous Waste, and UST regulations. Delaware Dept. of Natural Resources and Envtl. Control v. Nanticoke Homes, Inc., No. 98M-02-072 JOH (Del. Super. Ct. Feb. 12, 1998) (Consent Order). In addition to requiring the company to pay a $40,000 penalty to the DNREC, as well as $25,000 to a consultant to research its Delaware waste streams and develop stormwater management plans at two of its Delaware facilities, the consent order requires Nanticoke to implement an EMS and undertake a compliance audit and other actions at a cost of at least $175,000, pursuant to a related EMS Agreement. Under the EMS Agreement, Nanticoke must contract with an independent management consultant, who in turn will undertake an EMS Analysis, develop a Management Systems Improvement Plan, and perform a Compliance Audit. The EMS Analysis Workplan is to encompass: (i) criteria (e.g., evaluation of activities of company that are regulated under environmental law, and of systems with respect to requirements of applicable laws; recommendation of changes to or introduction of systems for effectively and efficiently managing activities with respect to compliance; recommendation of permanent systems for responding to changes in environmental regulations; preparation of a report); (ii) tasks (e.g., review or prepare supporting documentation; identify management systems; observe and evaluate actual operations; sample and test; interview facility personnel; review records and documents; assess pollution prevention potential; assess accident prevention and response procedures); (iii) schedule; (iv) personnel; and (v) DNREC comments and recommendations.

The consultant will prepare a Management Systems report describing the company's EMS, the findings of the EMS analysis, and any recommendations. The Management Systems Improvement Plan describes the actions the company has taken or will take to address the consultant's recommendations. The company must provide "a rationale and justification" for a decision not to adopt a recommendation of the consultant's report or a recommendation proposed by DNREC, and describe alternate measures the company has taken or intends to take to address the problem, or explain why the company disagrees that the underlying problem exists. The Compliance Audit includes: (i) a Compliance Audit Workplan; (ii) specific tasks; (iii) Compliance Audit Report; and (iv) DNREC comments and recommendations. In addition, Nanticoke was obligated to hire a full-time environmental compliance officer. The agreement does not require ISO 14001 certification or an audit of the EMS, and of the $175,000, only as much as $25,000 may be expended internally; the remainder must be paid to the outside EMS consultant.

26. A corporate policy statement regarding compliance with all environmental regulations must be included as a permanent part of the employee policy handbook distributed to all employees.

27. The company must appoint a company-wide environmental compliance officer at the executive level charged with the responsibility for assuring environmental compliance.

28. The agreement requires regular, company-wide periodic environmental compliance audits by qualified employees.

29. The company must establish a direct line of communication to the executive level compliance officer allowing any employee to contact the officer directly about possible environmental violations or with other questions concerning company environmental compliance.

30. The agreement mandates institution of company-wide policies to discipline employees for non-compliance with environmental regulations, and to recognize employees for actions taken to improve the company's environmental compliance and performance.

31. The agreement calls for establishment of a company-wide formal environmental compliance training program with annual updated environmental compliance training. The program must include both environmental compliance awareness training and job-specific environmental compliance training.

32. Reliance Electric-Highland, a division of Rockwell Automation, negotiated an EMS agreement with Indiana Department of Environmental Management ("IDEM") to settle alleged violations of federal (see 40 C.F.R. §§262.11 and 262.34, 40 C.F.R. Part 264, and 40 C.F.R. §268.50(a)(1)) and state (see,

NOTES TO APPENDIX E (CONT'D)

e.g., Ind. Code §13-30-2-1; IND. ADMIN. CODE tit. 329, r. 3.1-1-10, r. 3.1-15-4 and r. 3.1-15-8) hazardous waste laws and regulations. The January 1998 agreed order requires the company to pay a civil penalty of $13,922.25. Of this amount, $6,247.25 was to be paid in cash. State of Indiana v. Reliance-Electric Highland, Cause No. H-13028 (Jan. 16, 1998) (Agreed Order). In lieu of payment of the remainder, the company was obliged to perform and complete a SEP, costing a minimum of $30,700, and consisting of a commitment to obtain ISO 14001 certification within a specified timeframe. Significantly, under the SEP, the company must implement the elements of ISO 14001 by soliciting the input of a community-based stakeholder workgroup comprised of representatives from: (i) other local industries; (ii) local or regional environmental groups; (iii) community leaders; (iv) concerned citizens and neighbors; (v) local or state environmental agencies; and (vi) higher education institutions. The company was "strongly encouraged," though not required, "to share information related to local, state, or federal enforcement actions." The corporate EMS must identify potential opportunities or ideas for multimedia regulatory flexibility and convey such potential opportunities or ideas to the workgroup and IDEM. The company must submit progress reports quarterly until it is certified to ISO 14001. It must also complete a Final ISO 14001 Implementation and Certification Report attaching completed data protocols and addressing such questions as: (i) How has the ISO 14001 EMS helped to identify opportunities that go beyond compliance?; (ii) Has implementing ISO 14001 created better opportunities to introduce pollution prevention methodology into everyday business decisions?; (iii) How do you structure a stakeholder workgroup to develop effective communication between a facility and community?; (iv) How does implementation of an EMS affect relations between a facility and its customers and suppliers?; (v) How does implementation of an EMS affect inventory and operational control, as well as maintenance budgets?; (vi) After implementation, what priority is given to environmental concerns in everyday and long-term business decisions; (vii) How does an EMS affect compliance with respect to safety, fire hazards/protection and transportation regulations?

33. Reliance committed to develop and implement a comprehensive ISO 14001 environmental management system and to become ISO 14001 certified. It would be solely responsible for the proper design and implementation of the ISO 14001 environmental management system Pilot SEP.

34. The company is required to create an effective organizational structure and assign responsibility to its employees.

35. The company must develop an environmental management program; identify environmental aspects, legal and other applicable requirements; develop an environmental management system audit program; and, institute management review of the environmental management system.

36. The agreement calls for development of mechanisms for both internal and external communications.

37. The company must develop a monitoring and measuring program; institute procedures for operational control; and, implement a non-conformance and corrective/preventive action plan.

38. The agreement calls for solicitation of input from a community-based stakeholder workgroup consisting of local industries; representatives from local or regional environmental groups; community leaders; concerned citizens and neighbors; local or state environmental agencies; and, higher education institutions. The company agreed to organize the workgroup and hold a meeting at least once during the ISO 14000 process in order to discuss and critique the ISO implementation process. This workgroup would have an opportunity to tour the facility and make recommendations on the company's EMS. As "necessary," the workgroup could also share information with other industries and trade associations. The agreement also calls for the company to share information related to the implementation of its ISO 140001 Pilot SEP.

39. Donsco, Inc., charged by the Pennsylvania Department of Environmental Protection of violating the Pennsylvania Air Pollution Control Act, the Pennsylvania Clean Streams Law, the Pennsylvania Solid Waste Management Act, the Pennsylvania Storage Tank and Spill Prevention Act and the Pennsylvania Administrative Code, in April 1998 agreed to an EMS with multiple requirements. In re Donsco, Inc., Consent Order and Agreement with the Pennsylvania Department of Environmental Protection (Apr. 15, 1998). First, Donsco was obligated to hire a Corporate Environmental Systems Manager to develop an EMS Plan, and to retain a consultant to help complete the Plan. The company would then be obligated to submit an EMS Plan, developed on the basis of ISO 14001, for Department approval. The EMS Plan must encompass: (i) a corporate environmental policy statement reciting a commitment to environmental compliance, pollution prevention, and continued waste minimization; (ii) a source assessment of the manufacturing process and operations at the company's Pennsylvania plants for all sources of waste generation and identification of all discharges to air, water; (iii) an assessment of all pollution prevention opportunities for all sources of waste generation; (iv) an evaluation of the technical and economic feasibility of each pollution prevention opportunity; (v) establishment of specific pollution prevention objectives and targets; and (vi) establishment of an Environmental Action Program to include written procedures, documentation, employee training and schedules to achieve pollution prevention objectives and targets. Donsco had already decided to seek ISO 14001 certification. The Department proposed, and the company agreed, to implement an overall environmental management plan after submitting it to the Department for its approval.

40. General Motors Corporation was charged by DNREC with allegedly violating Delaware Air Pollution Control Regulations. In June 1998, the company was required to adopt the following EMS Requirements: to issue a written, plant-specific statement encompassing environmental performance and regulatory compliance; to obtain ISO 14001 certification by the year 2000; and to review and consider the use of EPA's Data Protocols for Environmental Management Systems. In re General Motors Corp., Delaware Department of Natural Resources and Environmental Control Secretary's Final Findings and Order (June 18, 1998). GM is under no obligation to incorporate these protocols as part of their ISO 14001 program.

41. Hussey Copper reached a similar agreement with the Pennsylvania Department of Environmental Protection ("DEP") in November 1997 settling alleged violations of the Pennsylvania Solid Waste Management Act, the Pennsylvania Clean Streams Law and the Pennsylvania Administrative Code. In re Hussey Copper Ltd., Consent Order and Agreement with the Pennsylvania Department of Environmental Protection (Nov. 17, 1997). To implement its EMS, the company first had to contract with a consultant to assist with the development of an EMS plan for the facility and submit an EMS plan for Departmental approval. Hussey's EMS plan was to include: (i) a corporate environmental policy reciting a commitment to environmental compliance, pollution prevention and continued waste minimization; (ii) a comprehensive source assessment of the manufacturing process and operations at the facility for all sources of waste generation and the identification of all discharges to air, water, or land; (iii) an assessment of the pollution prevention opportunities for all sources of waste generation; (iv) an evaluation of the technical and economic feasibility of each prevention opportunity; (v) establishment of specific pollution prevention objectives and targets; (vi) establishment of an Environmental Action Program containing written procedures, documentation, employee training and schedules to achieve pollution prevention objectives and targets; (vii) a schedule of implementation for the pollution prevention objectives and targets; and (viii) a schedule for management's review of the results of the EMS. Notably, the consent order defined "pollution prevention" as "source reduction and other practices that reduce or eliminate the creation of pollutants through increased efficiency in the use of raw materials, energy, water or other resources by conservation."

42. The order requires that General Motors ("GM") Wilmington obtain ISO 14001 certification.

NOTES TO APPENDIX E (CONT'D)

43. Donsco shall submit a plan approval application to modify the Air Permit; Individual Storm Water Permit Application for its discharge of stormwater from the facility; stormwater management plan providing a schedule for the removal stormwater drains adjacent to major process areas where spills can readily occur; and, a revised Preparedness, Prevention, and Contingency Plan which shall incorporate tank management procedures and provisions for decontaminating equipment used in the tank containment area and tank, and tank management.

44. The order obligates GM Wilmington to issue a plant-specific environmental statement within 12 months after the effective date of the decree and annually thereafter for 2 two years. This statement is to be in written form and provide an overview of environmental performance and regulatory compliance for the Wilmington Assembly Plant. The statement must include on-site releases to the environment pursuant to SARA Title III, the date reported in accordance with the Pollution Prevention Act, pollution prevention initiatives, and any violations or items of non-compliance at the GM Wilmington Plant that are cited by EPA, the DNREC, or any other violations which are required to be reported by law, as well as measures undertaken by the plant to correct them.

45. The company must state its commitment to environmental compliance, pollution prevention, and continued waste minimization.

46. Personnel must receive and respond to questions regarding environmental or regulatory compliance at the facility.

47. As noted in the consent order, the term EMS means that part of Hussey's overall management is responsible for maintaining the company's environmental policy, including the organizational structure and planning activities associated with developing and implementing the environmental policy.

48. Pursuant to Donsco's application to modify the Air Permit, the consent decree requires Donsco to complete the installation and startup of the New Furnace and the New Control System within 24 months of the date the Department completes its review and issues its final decision of application.

49. The company must enter into a contract with an environmental consultant to assist in developing an EMS plan for the facility.

50. On or before 30 days from the execution date of the consent order, Donsco must submit a plan to address any Cupola, Cupola Cooling System, and Cupola Control System failures that may occur prior to the installation and startup of the New Furnace and New Control System. The plan shall include, but not be limited to: a description of the steps that will be taken to minimize emissions from the Cupola in the event of a failure; procedures for notification of the Department; development of an employee training and awareness plan; and, a comprehensive preventive maintenance plan.

51. The agreement calls for an evaluation of the technical and economic feasibility of each pollution prevention opportunity; establishment and schedule of implementation of pollution prevention objectives and targets; and, establishment of an Environmental Action Program which shall include schedules to achieve pollution prevention objectives and targets.

52. As stipulated in the consent order, Donsco must make a set payment to the Commonwealth of Pennsylvania Solid Waste Abatement Fund, the Commonwealth of Pennsylvania Clean Water Fund and the Commonwealth of Pennsylvania Clean Air Fund.

53. The plant-specific environmental statement must be published in the Wilmington News Journal.

54. As stipulated in the consent order, Hussey must make a set payment to the Commonwealth of Pennsylvania Solid Waste Abatement Fund and the Commonwealth of Pennsylvania Clean Water Fund.

55. In July 1998, Keystone Cement Company and Pennsylvania Department of Environmental Protection settled, with specific EMS Requirements, alleged violations of the Pennsylvania Clean Streams Law, the Pennsylvania Solid Waste Management Act, the Pennsylvania Air Pollution Control Act, the Pennsylvania Administrative Code and the Pennsylvania Environmental Quality Board Regulations. In re Keystone Cement Co., Consent Order and Agreement with the Pennsylvania Department of Environmental Protection (July 10, 1998). First, Keystone must contract with an approved consultant, who will assist the company in developing an EMS. Subsequently, a schedule with milestones for the development and implementation of the EMS is to be submitted to the court. The EMS must address the following elements: (i) an environmental policy statement describing the commitment to continual improvement of the company's environmental performance, pollution prevention and compliance with applicable laws and regulations; (ii) identification of the environmental impacts of the company's activities, products or services; (iii) identification of all applicable federal, state, and local laws and regulations that impact the company, and include a statement describing how these legal requirements are factored into the company's operations; (iv) a statement identifying the company's environmental objectives and targets, consistent with its policy and environmental impacts; (v) a description and schedule indicating how the company will translate environmental objectives and targets into concrete action; (vi) investment of the resources necessary to ensure effective implementation (and identify personnel with responsibilities under the EMS); (vii) a training program to ensure that all employees are capable of performing their environmental responsibilities; (viii) a process governing communications with internal and external stakeholders; (ix) maintenance of documentation that describes the EMS, with provision for updates, as necessary; (x) a description of activities and processes where potential environmental impacts could be significant and documented procedures for those activities to avoid deviation from the environmental objectives and targets; (xi) identification of potential emergencies and development of procedures for preventing and responding to them; and (xii) a description of how the company will monitor and measure its EMS, and measures to be taken to correct deficiencies that may occur. There is no mention of ISO 14001 in the agreement.

56. Keystone must identify potential emergencies and develop procedures for preventing and responding to said emergencies. The company must also describe how it will monitor and measure its EMS, and in the event that deficiencies are identified with the EMS, Keystone must describe measures to be taken to correct such deficiencies.

57. The company must establish and maintain an environmental management system consistent with ISO 14001.

58. The company agreed to pay stipulated amount to the Toronto and Region Conservation Authority for the aid of the environmental education and awareness of children residents in its area of jurisdiction, in particular: "The Yellow Fish Road," a community outreach program that helps children protect community streams and educate their neighbors, and "Watershed on Wheels," an environmental conservation education program that travels into the community.

Consent Decree (U.S. v. ASARCO Inc.)

LOIS J. SCHIFFER
Assistant Attorney General
Environment and Natural Resources Division
United States Department of Justice

MICHAEL D. GOODSTEIN
Senior Attorney
Environmental Enforcement Section
Environment and Natural Resources Division
United States Department of Justice
P.O. Box 7611
Washington, D.C. 20044-7611
(202) 514-1111

JOHN N. MOSCATO
Trial Attorney
Environmental Enforcement Section
Environment and Natural Resources
999 18th Street, Suite 945
Denver, Colorado 80202
(303) 312-7346

SHERRY SCHEEL MATTEUCCI
United States Attorney
District of Montana

LORRAINE D. GALLINGER
Assistant United States Attorney
District of Montana
P.O. Box 1478
Billings, Montana 59103
(406) 657-6101

Attorneys for the United States

UNITED STATES DISTRICT COURT
FOR THE DISTRICT OF MONTANA

UNITED STATES OF AMERICA,	)	CONSENT DECREE
Plaintiff	)	Civil Action No:
v.	)	CV 98-3-H-CCL
ASARCO INCORPORATED,	)	
Defendant.	)	

[MATERIAL OMITTED]

I. BACKGROUND

WHEREAS, the United States and ASARCO have been engaged in national negotiations to resolve major environmental compliance issues at ASARCO facilities in a cooperative, innovative manner, without the transaction costs associated with protracted litigation;

WHEREAS, as a result of Phase I of these national negotiations, an agreement has been reached embodied in this consent decree and a related consent decree in the District of Arizona which provides for resolution of alleged claims by the United States against ASARCO under the Federal Resource Conservation and Recovery Act and the Federal Clean Water Act;

WHEREAS, the national agreement includes implementation of compliance measures, corrective action, and a supplemental environmental project at the East Helena Lead Smelter Facility, environmental protection and management measures at ASARCO's facilities nation-wide, and payment of civil penalties for alleged past violations at the East Helena Lead Smelter Facility under this consent decree; and, implementation of an extensive compliance plan at ASARCO's Ray Mine in Arizona and payment of civil penalties for alleged past violations under the decree in the District of Arizona;

WHEREAS, as the national agreement is embodied in two consent decrees the United States reserves the right to withdraw its consent from this Decree based on public comment received on either decree;

WHEREAS, concurrently with the lodging of this Consent Decree, Plaintiff, the United States of America ("United States"), on behalf of the United States Environmental Protection Agency ("EPA"), is filing a complaint (the "Complaint") against Defendant, ASARCO Incorporated ("ASARCO");

WHEREAS, ASARCO owns and operates a lead smelting facility in East Helena, Montana ("East Helena Facility" or "the Facility")

WHEREAS, the United States' claims are brought pursuant to the Resource Conservation and Recovery Act, as amended, 42 U.S.C. §§6901 et seq. ("RCRA") and the Federal Clean Water Act ("CWA"), 33 U.S.C. §1301 et seq., and seek the imposition of civil penalties and injunctive relief pursuant to the authority of Section 3008(a), (g), and (h) of RCRA, 42 U.S.C. §§6928(a), (g), and (h), and Section 309 (b) and (d) of the CWA, 33 U.S.C. §1319(b) and (d);

WHEREAS, ASARCO disputes the allegations of the Complaint, and its assent to this Consent Decree shall not constitute or be construed as an admission of liability; and

WHEREAS, the parties agree and the Court, by entering this Decree, finds that settlement of this matter, without protracted litigation, is fair, reasonable, and in the public interest.

NOW THEREFORE, it is hereby ORDERED, ADJUDGED, and DECREED as follows:

[MATERIAL OMITTED]

V. DEFINITIONS

8. Unless otherwise defined herein, terms used in this Decree shall have the meaning given to those terms in RCRA, the regulations promulgated thereunder, or the CWA, and the regulations promulgated thereunder.

Acceptable shall mean that the quality of submittals or completed work is sufficient to warrant EPA review in order to determine whether the submittal or work meets the terms and conditions of this Decree, including attachments, approved work plans and/or EPA's written comments and guidance documents. Acceptability of submittals or work, however, does not necessarily imply that they will be approvable. Approval by EPA of submittals or work, however, establishes that those submittals were prepared, or work was completed, in a manner acceptable to EPA.

Additional Work shall mean any activity or requirement that is determined by EPA to be necessary to fulfill the purposes of Part VII of this Decree, pursuant to the Additional Work provisions therein.

Administrative Record shall mean the record compiled and maintained by EPA in connection with the implementation of Part VII of this Decree.

AMS Standard shall mean, for purposes of this Decree only, the ASARCO Management System Standard dated November 17, 1997 (and any future revisions thereto - which affect ASARCO's obligations under this Decree as identified and described more fully in Paragraphs 152-174 below - upon mutual agreement of the parties and approval of the Court), which contains corporate-wide management system standards for key business objectives, including environmental, a copy of which is attached as Exhibit 4. This shall not be construed to prevent ASARCO from modifying its AMS Standard for all other purposes.

ANPR or Advance Notice of Proposed Rulemaking shall mean the ANPR published in the Federal Register on May 1, 1996 (61 Fed. Reg. 19432), which describes the evolution and present state of the corrective action program under RCRA, and also requests comment on the future direction of the corrective action program.

Areas to be Investigated as used in Part VII of this Decree shall mean any area of the Facility under the control or ownership of ASARCO where a release to the environment of any hazardous waste or hazardous constituent has occurred, is suspected to have occurred, or may occur, regardless of the time, frequency or duration of the release. The term Areas to be Investigated includes areas and/or discernible units at which solid wastes have been placed, at any time, irrespective of whether the area or unit was intended for the management of solid or hazardous waste. Examples of Areas to be Investigated include, but are not limited to landfills, surface impoundments, pits, waste piles, land treatment units, incinerators, tank systems (including any storage, treatment, or accumulation tank system), container storage units, elementary neutralization units, waste or wastewater treatment system units, transfer stations, and recycling units, or other areas or systems that received solid or hazardous waste or hazardous constituents, or released hazardous waste or hazardous constituents at any time. Investigations shall be designed so as to not inappropriately intrude into production operations, and to otherwise minimize any interference with production operations. All investigations shall be designed to support a determination of whether hazardous constituents have been, and/or are being released to the environment at, or from the area to be investigated.

ASARCO Facility(ies) shall mean unit(s) and complex(es) listed in Exhibit 3, which is incorporated by reference into this Consent Decree.

CAP or Final CAP or RCRA Corrective Action Plan shall mean the "RCRA Corrective Action Plan (Final)," OSWER Directive 9902.3-2A, dated May 1994.

CERCLA shall mean the Comprehensive Environmental Response, Compensation, and Liability Act of 1980, as amended, 42 U.S.C. §9601, et seq.

CERCLA Decree shall mean the Consent Decree entered into by EPA and ASARCO, and entered by the Federal District Court, District of Montana, Missoula Division, on December 27, 1990, as modified as of the date of entry of this Decree.

Comply or Compliance may be used interchangeably and shall mean completion of work required by Part VII of this Decree including submittal of documents of a quality acceptable to EPA, in accordance with work plans approved by EPA and in the manner and time specified in an approved work plan, Part VII of this Decree or any modification thereof. ASARCO must meet both the quality (see definition of acceptable) and timeliness components of a particular requirement to be considered to be in compliance with the terms and conditions of Part VII of this Decree.

Contractor shall include any consultant, laboratory and subcontractor retained to conduct or monitor any portion of the work performed pursuant to this Decree.

Corrective Measures shall mean those measures or actions appropriate to remediate, control, prevent, or mitigate the release, potential release or movement of hazardous waste or hazardous constituents into the environment or within or from one media to another.

Corrective Measures Implementation or CMI shall mean those activities appropriate to initiate, monitor, maintain, and complete the remedies EPA has selected or may select pursuant to Part VII of this Decree.

Corrective Measures Study or CMS shall mean the investigation and evaluation of potential alternative remedies to protect human health and/or the environment from the release or potential release of hazardous waste or hazardous constituents, into the environment from and/or at the Facility pursuant to Part VII of this Decree.

Day shall mean a calendar day unless expressly stated to be a business day. Business day shall mean a day other than a Saturday, Sunday, or Federal Holiday. In computing any period of time under this Decree, where the last day would fall on a Saturday, Sunday, or Federal Holiday, the period shall run until the end of the next Business day.

Decree shall mean this Consent Decree if, and as, entered by the Court, and any legally made modifications hereto.

DOJ shall mean United States Department of Justice Environmental Enforcement Section.

Effective Date shall be the effective date of this Decree.

EMS Policy shall mean, for purposes of this Decree only, the Environmental Management System Policy dated December 12, 1997 (and any future revisions thereto - which affect ASARCO's obligations under this Decree as identified and described more fully in Paragraphs 152-174 below - upon mutual agreement of the parties and approval of the Court), which contains requirements for the environmental component of the AMS Standard, a copy of which is attached as Exhibit 5. This shall not be construed to prevent ASARCO from modifying its EMS Policy for all other purposes.

Encycle Materials shall mean materials received at the East Helena Facility from the Encycle/Texas facility in Corpus Christi, Texas through the date of lodging of this Decree.

Environmental Aspects shall mean element of ASARCO's activities, products, or services which can interact with the environment

Environmental Impacts shall mean any change to the environment, whether adverse or beneficial, wholly or partly resulting from ASARCO's activities, products, or services.

Environmental Requirements shall mean all applicable federal, state, and local environmental statutes and regulations, including permits and enforceable agreements between ASARCO and the respective environmental regulatory agencies.

EPA shall mean the United States Environmental Protection Agency, and any successor departments or agencies of the United States.

Facility as used in the term "the Facility" or in the term "East Helena Facility" shall mean ASARCO's East Helena primary lead smelter and contiguous areas under ASARCO's ownership and control at East Helena.

Ground Water shall mean the water in the saturated zone beneath the land surface.

Hazardous Constituents shall mean those constituents listed in Appendix VIII to 40 C.F.R. Part 261 or any constituent identified in Appendix IX to 40 C.F.R. Part 264 or any approved subset of Appendix IX to 40 C.F.R. Part 264.

Hazardous Waste shall mean hazardous waste as defined in section 1004(5) of RCRA, 42 U.S.C. §6903(5), and 40 C.F.R. §261.3.

Imminent Threat shall mean any release, or threatened release, of hazardous waste or a hazardous constituent, at or from the Facility, which may present an imminent or substantial endangerment to human health and/or the environment.

Interim Measure or IM shall mean those actions which can be, or are, initiated in advance of implementation of the final Corrective Measures for the Facility.

On-Site Contractor shall mean ASARCO contractors and subcontractors that have fixed-base operations at ASARCO Facilities for more than thirty (30) days in a calendar year, including, but not limited to, raw material blending, residuals processing, vehicle maintenance, and/or waste treatment.

Operating Smelting Facilities shall mean ASARCO's smelter facilities at East Helena, Montana; El Paso, Texas; Hayden, Arizona; Glover, Missouri, and any such additional facilities which commence operations during the pendency of this Decree.

Paragraph(s) shall mean a paragraph or paragraphs of this Decree numbered with arabic numerals, including all subparagraphs contained therein.

Part of this Decree shall mean a portion of this Decree identified by an upper case roman numeral.

RCRA shall mean the Solid Waste Disposal Act, as amended by the Resource Conservation and Recovery Act, as further amended by, inter alia, the Hazardous and Solid Waste Amendments of 1984, and the Federal Facility Compliance Act of 1992, 42 U.S.C. §6901 et seq.

RCRA Facility Investigation or RFI shall mean the investigation and characterization of the source(s) and/or releases of hazardous wastes and/or hazardous constituents and the nature, extent, direction, rate, movement, and concentration of such releases of hazardous wastes and/or hazardous constituents, that have been, or may be released or may reasonably be expected to be released into the environment from or at, and/or to migrate from the Facility.

Receptors shall mean those animals (including humans) or plants, and their habitats which are or may receive or be affected by releases of hazardous waste or hazardous constituents at, or from, the Facility.

Record of Decision or ROD shall mean the Record of Decision issued by EPA Region VIII in November, 1989, regarding the East Helena Facility, and as modified through the date of entry of this Decree.

Release shall mean any spilling, leaking, pumping, pouring, emitting, emptying, discharging, injecting, escaping, seeping, leaching, dumping, placing, or disposing into the environment.

Risk Assessment shall mean the process or framework for characterizing risk to human and ecological receptors.

Risk Assessment Guidance or RAGS shall mean the Risk Assessment Guidance for Superfund (December 1989).

Secondary Materials shall mean materials that potentially can be a solid and hazardous waste when recycled. The following are considered secondary materials: spent materials, sludges, by-products, scrap metal, and commercial chemical products recycled in ways that differ from their normal use. Each of these types of secondary materials is defined at 40 C.F.R. §261.1.

Submittal shall include any work plan, report, progress report, or any other written document ASARCO is required by this Decree to send to EPA.

Violation(s) shall mean actions, omissions, failures, or refusals to act by ASARCO that result in a failure to meet the terms and conditions of this Decree.

Work shall mean any activity ASARCO must perform to comply with the requirements of this Decree.

Work Plans shall mean the detailed plans prepared by ASARCO to satisfy the requirements of each phase of work under Part VII of this Decree. The requirements for each work plan are set forth in the section of Part VII of this Decree entitled Work To Be Performed. All work plans and modifications or amendments thereto, shall be filed with the Court to be incorporated into this Decree and are an enforceable component of this Decree when approved in writing by EPA, unless otherwise required under Part VII of this Decree.

[MATERIAL OMITTED]

X. ENVIRONMENTAL MANAGEMENT AND PROTECTION REQUIREMENTS

152. The purpose of this Part of the Decree is to enhance ASARCO's development of a comprehensive corporate-level environmental management system to promote compliance with all environmental requirements, achieve pollution prevention, and accomplish pollution reduction. To this end, ASARCO has developed corporate-wide management system standards, which are documented in the AMS Standard and an EMS Policy which put forth requirements for the Environmental Management System.

153. For purposes of this Decree, ASARCO shall develop and implement those portions of the AMS Standard and EMS Policy identified and more fully described in Paragraphs 154 through 174 below. The AMS Standard and EMS Policy are appended to and incorporated by reference into this Decree as Exhibits 4 and 5. All numeric references included in Paragraphs 154 through 174 refer to specific provisions of the AMS Standard or EMS Policy. The provisions of Paragraphs 154 through 174 and any related exhibits apply to ASARCO Facilities listed in Exhibit 3.

ENVIRONMENTAL COMPLIANCE PROMOTION

Employee Environmental Awareness Training

154. ASARCO shall develop and implement an Employee Environmental Awareness Training Program as internally mandated by the AMS Standard and EMS Policy. This program is intended, in part, to increase employee awareness and involvement in meeting an ASARCO Facility's Environmental Requirements. Specifically, ASARCO shall initiate implementation of the applicable requirements of AMS Standard 900 and the EMS Policy Section 4.3.2, including the activities described in this subsection, at all ASARCO Facilities, within twelve (12) months after this Decree is entered by the Court.

a. ASARCO shall develop and conduct environmental awareness training for all employees at all ASARCO Facilities. The training shall include at least four (4) hours of initial classroom instruction, which may include interactive computer-based instruction, and an annual refresher of at least two (2) hours of classroom instruction. Initial classroom instruction for all existing employees shall be conducted within twenty-four (24) months after this Consent Decree is entered by the Court.

b. ASARCO shall also develop a program to ensure that all new employees are provided training equivalent to the four-hour initial classroom instruction within three (3) months after they begin work at any of the ASARCO Facilities. An employee transferred from another ASARCO Facility or a rehired employee who has already received the initial training or equivalent shall not be considered a new employee for the purpose of this requirement.

c. ASARCO shall submit a generic training plan in accordance with subparagraph a. above, to EPA for review and comment within six (6) months after this Decree is entered by the Court. EPA may submit any comments to be considered by ASARCO within sixty (60) days of submission of the generic training plan by ASARCO. The training plan shall, as a minimum, include:

(1) Prototype training modules for the topics identified in the following subparagraph.
(2) List of trainers (by position title), and their individual qualifications and experience.
(3) Schedule for completing the initial training at each ASARCO Facility.
(4) A description of the environmental awareness component of the new employee training program.
(5) A description of how the training will be reinforced in the workplace in order to maintain increased employee awareness and involvement in meeting the facility's environmental requirements.
(6) A periodic (minimum of at least one annually) evaluation of the training program and how ASARCO plans to incorporate the results into program improvements.

d. Subject material for the initial classroom component of each training program shall be documented in a training manual and address, at a minimum, the following elements and all elements contained in the EMS Policy, Section 4.3.2.:

(1) Training Objectives
(2) ASARCO Environmental, Safety and Health Policy
(3) Overview of Pollution and Potential Environmental Impacts of ASARCO Facility Operations
(4) Environmental Regulatory Requirements (Statute Overview)
(5) Hazards and the Environment
(6) Facility-Specific Operations, Waste Generation and Management, and Environmental Requirements
(7) Emergency Preparedness and Response
(8) Role of Environmental Department
(9) Role of Individual Employees at the ASARCO Facility (including potential consequences of departure from specified operating procedures, such as non-compliance and/or pollutant releases to the environment)
(10) Pollution Prevention/Recycling/Waste Minimization Programs
(11) Community Outreach Activities

e. Subject material for the refresher classroom component of each training program shall be documented in a training manual and address, as a minimum, the following elements:

(1) Training Objectives
(2) ASARCO Environmental, Safety and Health Policy
(3) New or Revised Environmental Requirements
(4) Review of Employee Environmental Responsibilities (including potential consequences of departure from specified operating procedures)

Protocol For Identifying and Communicating Environmental Requirements

155. ASARCO shall develop and implement a corporate-level program, as more particularly described below, to identify and document environmental requirements applicable to ASARCO Facilities and to establish additional training requirements for company environmental professionals. Specifically, ASARCO shall develop and initiate implementation of the applicable elements of AMS Standard 200 for Customer Requirements, Standard 910 for Qualification and Training, and EMS Policy, Section 4.2.2 for Legal and Other Requirements, and EMS Policy, Section 4.3.2 for Training, Awareness, and Competence, and complete the requirements described in this Paragraph within twelve (12) months after this Decree is entered by the Court.

156. ASARCO shall document this corporate-level program for identifying, understanding, and communicating environmental requirements to affected ASARCO Facilities and other ASARCO personnel, service providers, and On-Site Contractors, and monitor facility activities for conformance to those requirements. The protocol shall also address identifying, obtaining, and evaluating information about changes and proposed changes in Environmental Requirements that could potentially apply to and/or impact company operations. The program plan shall be submitted to EPA for review and comment within six (6) months after this Decree is entered by the Court. EPA may submit any comments to be considered by ASARCO within sixty (60) days of submission of the plan by ASARCO. The plan shall describe, at a minimum, the following elements:

a. The process for identifying applicable and potentially applicable environmental requirements.

b. The communication protocol for disseminating information on applicable and/or potentially applicable Environmental Requirements to, as a minimum, the individuals included in the network described in subparagraph c. below.

c. The process for interpretation and assessment of applicability through an identified network of designated knowledgeable individuals, defined as those environmental professionals at the corporate level and at each affected ASARCO Facility, who are familiar with the company's production and ancillary processes, Environmental Requirements, and other aspects of operations that could affect the environment and/or compliance.

d. The implementation process at each ASARCO Facility, including methods to communicate the applicable requirements to service providers and On-Site Contractors in a timely manner.

157. The protocol described above shall also establish minimum qualifications and training requirements, including both core and refresher training, for members of the network of designated knowledgeable individuals (described in Paragraph 156 subparagraph c. above). The program plan shall describe measures to be implemented by ASARCO to ensure that network members satisfy the established qualification and training requirements on a continuing basis.

158. Finally, the protocol described above shall provide for a periodic (minimum of one annually) evaluation of the protocol and describe how ASARCO will incorporate the results into program improvements.

Technology Transfer

159. After the protocol described in Paragraph 155 above has been developed and implementation initiated, ASARCO shall share the protocol and lessons learned with other businesses and organizations through a series of seminars as more specifically discussed below in Paragraphs 160 and 161. These seminars shall focus on, at a minimum: (1) how potentially applicable current and proposed environmental requirements are systematically identified; (2) how these requirements are communicated, discussed, interpreted, and assessed for applicability to company operations and facilities; (3) how applicable requirements are integrated into ASARCO Facility activities and processes; (4) practical strategies for implementing this protocol at the plant level; and (5) discussion of implementation/operation successes and problems.

160. ASARCO shall provide or make arrangements for providing the facilities for such seminars, and shall provide all necessary publicity and invitations for them. ASARCO shall plan the curricula for these sessions. However, representatives from EPA and State and local environmental regulatory agencies may be invited to actively participate in the planning of these sessions, and may attend them. The seminar sessions may be led by, or include participation by, outside experts provided by ASARCO. These seminars shall be offered to the participants free of charge.

161. In each of the fourth and fifth years after this Decree is entered by the Court, ASARCO will conduct three seminars, as described in Paragraphs 159 and 160 above, in forums (e.g., meetings of national or regional professional organizations) that will be attended by members of the regulated community from Regions where ASARCO Facilities are located. Information about seminars conducted in fulfillment of these requirements shall be provided to EPA in the Progress Reports described in Paragraph 178 below.

POLLUTION PREVENTION AND POLLUTION REDUCTION

Environmental Management System Implementation

162. ASARCO shall fully implement the Environmental Objective of the ASARCO Management System in accordance with the AMS Standard and EMS Policy, at all ASARCO Facilities in accordance with the following provisions.

163. ASARCO shall develop an EMS Implementation Plan and submit the plan to EPA for review and comment within three (3) months after this Decree is entered by the Court. EPA may submit any comments to be considered by ASARCO within sixty (60) days of the submission of the implementation plan by ASARCO. The plan shall contain an implementation schedule with milestones for each ASARCO Facility. Further, each plan shall identify individuals (by position) who are responsible for EMS implementation at corporate and facility levels and their respective responsibilities and authorities. At a minimum, the milestones shall include:

a. Completion of an initial review and evaluation of the current EMS, as described in Paragraph 164 below.

b. Completion of initial plans for development of implementing documents and tasks (hereafter, Development Plan), as described in Paragraph 165 below.

c. Completion of the Final Implementation Report, as described in Paragraph 167 below.

164. In accordance with the schedule contained in the EMS Implementation Plan, ASARCO shall conduct an initial review and evaluation of the current EMS at each ASARCO Facility to identify and assess the potential impacts of program gaps relative to the AMS Standard and the EMS Policy for the purpose of prioritizing development of implementing documents and other tasks.

165. In accordance with the schedule contained in the EMS Implementation Plan, ASARCO shall complete the initial Development Plan for each ASARCO Facility. The Development Plans shall be based on the initial review and evaluation results and other information, and shall contain a document development schedule, and listing and scheduling of other implementing tasks. Each specific document and task identified in the Development Plan shall be cross-referenced to the respective AMS Standard and EMS Policy section, where appropriate. The format for each Development Plan shall be the same and shall be based on the format presented in Exhibit 6. Within thirty (30) days following completion of each plan, but not later than twelve (12) months after this Decree is entered by the Court, ASARCO shall submit each Development Plan to EPA for

review and comment. EPA may submit any comments to be considered by ASARCO within sixty (60) days of the submission of the development plan by ASARCO. The submittal shall be in both paper hard copy and a mutually agreeable electronic format.

166. Within fifteen (15) months after entry of this Decree, ASARCO shall initiate implementation of the Development Plans in accordance with the schedule of implementing tasks required by Paragraph 165 above.

167. In accordance with the schedule contained in the EMS Implementation Plan, ASARCO shall submit a Final Implementation Report to EPA for each ASARCO Facility that shows completed Development Plan elements, and both scheduled and actual completion dates. The submittal shall be in both paper hard copy and a mutually agreeable electronic format ASARCO shall complete implementation of the Development Plans at all ASARCO Facilities not later than forty-eight (48) months after this Decree is entered by the Court. Included in the Final Implementation Report will be a compilation of the data collected pursuant to Paragraph 174 below.

Environmental Aspects/Environmental Impacts Review

168. ASARCO shall conduct an expedited Environmental Aspects/ Environmental Impacts review at all ASARCO Facilities. Based on that review and other pertinent information, ASARCO shall develop and implement an assessment, prevention, and control (APC) program, as more particularly described below. Program implementation shall include development of detailed process flow diagrams for each operation in each ASARCO Facility, as described below. ASARCO's APC Program shall include, at a minimum, processes for:

a. Ongoing assessments of operations, for the purposes of preventing and controlling releases, environmental protection, and maintaining compliance with statutory and regulatory requirements. These assessments shall include the following activities:

(1) monitoring and measurements, as appropriate, to maintain compliance (Refer to EMS Policy, Section 4.4.1);
(2) identifying operations and waste streams where equipment malfunctions and deterioration, operator errors, and discharges or emissions may be causing, or may lead to, releases of hazardous waste or hazardous constituents to the environment, or a threat to human health or the environment (Refer to EMS Policy, Section 4.3.6); and
(3) performing root cause analysis of identified problems to prevent recurring issues.

b. Identifying activities that could cause adverse environmental impacts and/or regulatory non-compliance and, where documented, standard operating practices need to be developed (Refer to EMS Policy, Section 4.3.6).
c. Conducting and documenting routine, objective, self-inspections by department supervision and trained staff, especially at locations identified by the process described in subparagraph a. (2) above (Refer to EMS Policy, Section 4.4.1).
d. Ensuring input of environmental concerns and requirements in planning, design, and operation of ongoing, new, and/or changing buildings, processes, maintenance activities, and products (Refer to EMS Policy, Sections 4.2.1 and 4.2.4).
e. Developing and implementing corrective action procedures which address the following internal/external indicators of actual and/or potential non-compliance, if they occur:

(1) notice from a government agency of alleged non-compliance;
(2) observed or documented instance of noncompliance with regulatory or permit requirements;
(3) observed or documented instance of noncompliance with EMS policies or procedures;
(4) citizens' complaints; or
(5) significant environmental impact resulting from activities enumerated in subparagraphs (1) - (4) above.

f. For the purposes of this section, corrective action procedures will include, at a minimum, the following activities: identification and evaluation of any instance of actual and/or potential non-compliance; communication of noncompliance to appropriate operations management with responsibility for authorizing corrective action; determination of and approval for necessary corrective action; schedule for implementation, tracking and reporting of corrective action taken, as needed; and actual implementation and follow-up of corrective action (Refer to EMS Policy, Section 4.2.4).

169. ASARCO shall develop a plan and schedule that describes how, when, and by whom (by position title and qualifications) the environmental aspects/impacts review will be conducted at each ASARCO Facility. The plan shall be submitted to EPA for review and comment within three (3) months after this Decree is entered by the Court. EPA may submit any comments to be considered by ASARCO within sixty (60) days of the submission of the plan by ASARCO. ASARCO shall ensure that reviews at the facilities listed in Exhibit 3, Group 1 are conducted as a first priority. All environmental aspects/impacts reviews shall be completed within eight (8) months after this Decree is entered by the Court.

170. An APC Program is internally mandated by the EMS Policy and the Environmental Objective of the AMS Standard. ASARCO shall develop and complete implementation of the applicable APC requirements of AMS Section 4 Standards for Process Management and EMS Policy Sections referenced in Paragraph 168 a - f above, and incorporate the APC processes identified in Paragraph 168 above, at all ASARCO Facilities, within thirty (30) months after this Decree is entered by the Court.

171. ASARCO shall develop detailed process flow diagrams and associated descriptive information for each ASARCO

and On-Site Contractor operation in each ASARCO Facility, as follows:

a. Identify each process and/or process group and prepare a narrative describing each such operation, including process steps, production capacity and the most recent annual production in a hard copy or electronic "fact sheet."
b. Develop a schematic process flow diagram for each current process and/or process group. All process flow diagrams developed pursuant to this requirement shall use consistently defined terms and labels. A glossary of commonly used terms shall be prepared and updated, as necessary. Each process flow diagram shall, as a minimum, depict each unit operation and identify (by name) raw materials, products, by-products, recyclable materials and waste streams.
c. On each process flow diagram, for each waste stream, by-product, and recyclable material exiting the process, identify its final disposition. Identify any state-assigned emission point source identification number for air emission sources and indicate the respective National Pollutant Discharge Elimination System ("NPDES") permitted outfall for wastewater streams.
d. For each process or process group, indicate the presence of side sumps, pump stations, and hazardous waste accumulation areas containing 55 gallons or more.
e. If a hazardous waste determination was made for a waste stream or group of waste streams identified on the process flow diagram, indicate the results of that determination either on the schematic or on the "fact sheet" and quantity generated during the last calendar year.
f. For each ASARCO On-Site Contractor, the Environmental Department shall maintain pertinent information including complete company names, mailing addresses, principal contacts for environmental matters and their telephone numbers, RCRA I.D. numbers, if applicable, and brief description of the services rendered and include this information on the fact sheet.

172. ASARCO shall review annually and update as necessary the process flow diagrams and fact sheet information to incorporate new equipment, or processes, or modifications to existing equipment or processes. ASARCO shall include all revision dates on the process flow diagrams and fact sheets.

173. ASARCO shall submit a schedule for completing the process flow diagrams and associated fact sheets at each ASARCO Facility to EPA for review and comment within eight (8) months after this Decree is entered by the Court. EPA may submit any comments to be considered by ASARCO within sixty (60) days of the submission of the schedule by ASARCO. The schedule shall indicate higher priority for completion of this work at the ASARCO Facilities listed in Exhibit 3, Group 1; this work will be completed for all ASARCO Facilities within twenty-four (24) months after this Decree is entered by the Court. ASARCO shall also provide with the schedule, at least three (3) prototype process flow diagrams and associated fact sheets, each from a different ASARCO Facility.

Environmental Metrics

174. ASARCO shall collect and report data on the Environmental Metrics listed in Exhibit 7 hereto for each ASARCO Facility on an annual basis for the purpose of measuring the impacts of implementation of the EMS. ASARCO shall propose detailed monitoring parameters and reporting format in accordance with Exhibit 7 to EPA for review and comment within three (3) months after this Decree is entered by the Court. EPA may submit comments to be considered by ASARCO within sixty (60) days of the submission of the monitoring parameters and reporting format by ASARCO.

XI. REPORTING

Reporting: Materials Management

175. In accordance with Paragraph 12 of this Decree, ASARCO shall submit a report to EPA annually on any revisions made to the acceptance criteria in the preceding year with a written explanation of the underlying justification for the revisions. The annual report shall also specify the date on which ASARCO began following the revised procedures.

Reporting Corrective Action

176. The monthly progress reports deliverable pursuant to Part VII of this Decree are to be sent to EPA no later than the twentieth (20th) day of the following month and shall at a minimum:

a. Describe the actions, progress, and status of projects which have been undertaken pursuant to Part VII of this Decree;
b. Identify any requirements under Part VII of this Decree that were not completed in a timely manner, and problem or anticipated problem areas affecting compliance with this Decree;
c. Describe projects completed during the prior month, as well as the activities scheduled for the next month;
d. Describe, and estimate the percentage of, studies completed;
e. Describe and summarize all findings to date;
f. Describe actions being taken to address problems;
g. Identify changes in key personnel during the period;

h. Include copies of the results of sampling and tests conducted and other data generated pursuant to work performed under Part VII of this Decree since the last Progress Report. ASARCO may also submit data that have been validated and confirmed by ASARCO to supplement any prior submitted data. Updated validated and confirmed data shall be included with the RFI Report if not delivered before; and

i. Describe the status of financial assurance mechanisms, including whether any changes have occurred, or are expected to occur which might affect them, and the status of efforts to bring such mechanisms back into compliance with the requirements of this Decree.

Reporting: East Helena SEP

177. In accordance with Part VIII, a report detailing baseline conditions shall be submitted to EPA within thirty (30) days after implementation of Phase 1 of the SEP. Annual reporting requirements prior to SEP completion shall consist of the preparation of yearly reports summarizing the findings of the vegetation, wildlife and soils sampling and monitoring. No later than sixty (60) days after ASARCO completes the SEP described in Paragraphs 144 - 146 above, ASARCO shall provide EPA a SEP Completion Report containing the following information: (i) A description of the SEP as fully implemented; (ii) A description of the environmental benefits resulting from implementation of each SEP; (iii) Itemized and total net costs of the SEP; and (iv) A certification that the SEP has been fully implemented pursuant to the provisions of this Consent Decree and the Statement of SEP Work.

Reporting: Environmental Management And Protection

178. ASARCO shall submit semi-annual progress reports to EPA summarizing progress made in completing the activities required by Paragraphs 154 through 173 above. Progress reports shall be submitted to EPA within thirty (30) days after the last day of June and December of each calendar year commencing in 1998 until all activities required by Paragraphs 154 through 173 above are completed. The progress reports shall include any and all revisions to the AMS Standard and EMS Policy made during the reporting period.

a. The progress reports, as appropriate, shall contain a summary of the employee environmental awareness training given, as required by Paragraph 154 above, and include the following information:

(1) A listing of where and when training classes were given.
(2) Certification that the required training was conducted by ASARCO pursuant to the provisions of this Decree.
(3) Costs of conducting this training during the reporting period.
(4) Number of employees trained during reporting period.

b. The progress reports, as appropriate, shall contain a summary of the seminars required by Paragraphs 159 through 161 above, and include the following information:

(1) A description of where and when the seminar was given and the audience.
(2) Certification that the required seminars were conducted by ASARCO pursuant to the provisions of this Decree.
(3) Costs of conducting the seminars during the reporting period.
(4) List of attendees.

c. The progress reports, as appropriate, shall contain a summary of the EMS implementation required by Paragraph 170 above, and include the following, as required by Paragraph 162 above:

(1) Total number of procedures requiring documentation in the EMS as of the last report.
(2) Number of procedures requiring documentation added during reporting period.
(3) Number of procedures no longer requiring documentation identified during reporting period.
(4) Total number of procedures requiring documentation as of the end of this reporting period.
(5) Number of procedures requiring documentation due during this reporting period.
(6) Number of procedures requiring documentation completed during reporting period.

d. The progress reports, as appropriate, shall describe progress made in completing the process flow diagrams and associated fact sheets for each ASARCO Facility, as required by Paragraph 171 above, and include the following information:

(1) Expected completion date for each ASARCO Facility.
(2) Change, if any in expected completion date from previous progress report.
(3) List of ASARCO Facilities for which process flow diagrams and associated fact sheets have been completed.

e. The progress reports due within thirty (30) days after the end of June, as appropriate, shall include the data collected during the previous calendar year for the Environmental Metrics, as required by Paragraph 174 above.

[MATERIAL OMITTED]

XXIV. TERMINATION

233. This Decree shall terminate upon motion of any party after each of the following has occurred:

a. ASARCO has completed all actions required by this Decree;
b. ASARCO has paid all monies and penalties due under this Decree;
c. ASARCO has submitted a certification to the United States that conditions a. and b., above, have been met and
d. Within 60 days of receiving such certification from ASARCO (or longer, if the United States notifies ASARCO that additional time will be required), the United States has not disputed, in writing, ASARCO's contention that conditions a. and b. above have been met. If the United States disputes ASARCO's contention, the dispute resolution provision of this Decree (Part XVI) shall be invoked and this Decree shall remain in effect pending resolution of the dispute by the parties or the Court.

Dated and entered this ____ day of

UNITED STATES DISTRICT JUDGE

WE HEREBY CONSENT to the entry of this Decree, subject to the public notice requirements of 28 C.F.R. §50.7.

FOR THE UNITED STATES OF AMERICA:

WE HEREBY CONSENT to the entry of this Decree, subject to the public notice requirements of 28 C.F.R. §50.7.

DATE

LOIS J. SCHIFFER
Assistant Attorney General
Environment and Natural Resources Division
United States Department of Justice

DATE

MICHAEL D. GOODSTEIN
Senior Attorney
Environmental Enforcement Section
Environment and Natural Resources Division
United States Department of Justice

DATE

JOHN N. MOSCATO
Trial Attorney
Environmental Enforcement Section
Environment and Natural Resources Division
United States Department of Justice

SHERRI SCHEEL MATTEUCCI
United States Attorney
District of Montana

DATE

LORRAINE D. GALLINGER
Assistant United States Attorney

WE HEREBY CONSENT to the entry of this Decree, subject to the public notice requirements of 28 C.F.R. §50.7.

DATE

STEVEN A. HERMAN
Assistant Administrator for Enforcement
U.S. Environmental Protection Agency
Washington, D.C.

WE HEREBY CONSENT to the entry of this Decree, subject to the public notice requirements of 28 C.F.R. §50.7.

DATE	William P. Yellowtail Regional Administrator U.S. Environmental Protection Agency, Region VIII
DATE	Carol Rushin Assistant Regional Administrator U.S. Environmental Protection Agency, Region VIII Office of Enforcement, Compliance and Environmental Justice
DATE	Michael T. Risner Director, Legal Enforcement Program U.S. Environmental Protection Agency, Region VIII Office of Enforcement, Compliance and Environmental Justice

WE HEREBY CONSENT to the entry of this Decree, subject to the public notice requirements of 28 C.F.R. §50.7.

FOR DEFENDANT ASARCO, INC.:

DATE	Name: Title:
DATE	Name: Title:

Consent Decree (U.S. and Texas v. Encycle/Texas, Inc. and ASARCO Inc.)

UNITED STATES DISTRICT COURT
SOUTHERN DISTRICT OF TEXAS

UNITED STATES OF AMERICA,		
and		
STATE OF TEXAS,	)	CONSENT DECREE
Plaintiffs,	)	
	)	
V	)	H –99-1136
ENCYCLE/TEXAS, INC.	)	
	)	
and		Civil Action No:
ASARCO INCORPORATED,	)	
Defendants.	)	

[MATERIAL OMITTED]

I. BACKGROUND

WHEREAS, Encycle/Texas, Inc. ("Encycle") owns and operates a facility in Corpus Christi, Texas ("Corpus Christi Facility" or "Encycle Facility")

WHEREAS, ASARCO Incorporated ("ASARCO") owns and operates facilities in El Paso, Texas ("El Paso Facility"), Amarillo, Texas ("Amarillo Facility"), and in East Helena, Montana ("East Helena Facility") and in Tennessee ("Tennessee Mines") ("the Facilities")

WHEREAS, the United States, the State of Texas ("State"), on behalf of the Texas Natural Resource Conservation Commission ("TNRCC"), and Encycle, have engaged in negotiations to resolve environmental compliance issues at the Encycle Facility;

WHEREAS, the United States and ASARCO have been engaged in national negotiations to resolve environmental compliance issues at ASARCO facilities in a cooperative, innovative manner, without the transaction costs associated with protracted litigation;

WHEREAS, concurrently with the lodging of this Consent Decree ("Consent Decree" or "Decree"), Plaintiffs, the United States of America ("United States"), on behalf of the United States Environmental Protection Agency ("EPA") and the State of Texas, (collectively "Plaintiffs") are filing a complaint (the "Complaint") against Defendants, Encycle and ASARCO as provided herein;

WHEREAS, the United States' claims are brought pursuant to the Resource Conservation and Recovery Act, as amended, 42 U.S.C. §§6901 et seq. ("RCRA") and the Federal Clean Water Act, 33 U.S.C. §1301 et seq. ("CWA"), and seek the imposition of civil penalties and injunctive relief pursuant to the authority of Section 3008(a), (g), and (h) of RCRA, 42 U.S.C. §§6928(a), (g), and (h), and Section 309(b) and (d) of the CWA, 33 U.S.C. §1319(b) and (d)

WHEREAS, the State's claims are brought pursuant to RCRA, and Chapter 361 of the Texas Health and Safety Code, Tex. Health & Safety Code Ann. §361.001 et seq. and Chapter 7 of the Texas Water Code, Tex. Water Code Ann. §7.001 and the rules and regulations promulgated pursuant to those statutes;

WHEREAS, Encycle and ASARCO dispute the allegations of the Complaint, and their assent to this Consent Decree shall not constitute or be construed as an admission of liability;

WHEREAS, an agreement has been reached embodied in this Consent Decree which provides for resolution of the RCRA, CWA and State law civil claims in the Complaint by the United States and Texas against ASARCO, and Encycle;

WHEREAS, the United States and the State (collectively "Plaintiffs"), and ASARCO and Encycle (collectively "Settlors") agree and the Court, by entering this Decree, finds that settlement of this matter, without protracted litigation, is fair, reasonable, and in the public interest.

NOW THEREFORE, it is hereby ORDERED, ADJUDGED, and DECREED as follows:

II. JURISDICTION AND VENUE

1. This Court has jurisdiction over the subject matter of this action and over the parties pursuant to Section 3008 (a) of RCRA, 42 U.S.C. §6928(a) and Section 309 of the CWA, 33 U.S.C. §1319, and 28 U.S.C. §§1331, 1345, 1355, and 1367.

2. Venue is proper in the Southern District of Texas pursuant to Section 3008(a) of RCRA, 42 U.S.C. §6928, Section 309(b) of the CWA, 33 U.S.C. §1319(b) and 28 U.S.C. §§1391 and 1395(a). The Complaint states a claim upon which relief may be granted.

III. BINDING EFFECT

3. This Decree shall apply to, and be binding upon Settlors and their successors and assigns, jointly and severally. Settlors certify that their undersigned representatives are fully authorized to enter into the terms and conditions of this Consent Decree, to execute it on behalf of ASARCO and Encycle, and to legally bind ASARCO and Encycle to its terms.

4. The Settlors agree to be bound by this Decree and not to contest its validity in any subsequent proceeding to implement or enforce its terms.

5. Except as provided in Paragraph 63, no change in ownership or corporate or other legal status including, but not limited to, any transfer of assets or property, shall alter Settlors' responsibilities under this Decree. No later than 30 days prior to sale or transfer of ownership or operation of the Corpus Christi, El Paso or Tennessee Mines Facilities, Settlors shall give written notice of this Decree to each purchaser or successor in interest. Upon such sale or transfer, Settlors shall provide a copy of this Decree to each such purchaser or successor in interest. The sale or transfer of ownership or operation of the Corpus Christi or El Paso Facility will not relieve Settlors of their obligations under this Decree, unless this Decree is modified pursuant to Part XXVII (Modification) to reflect such a change Settlors shall notify in writing the United States Department of Justice (DOJ), EPA, and the TNRCC of such purchaser or successor in interest at least 30 days prior to any sale or transfer.

6. Settlors shall provide a copy of this Decree to each contractor retained to perform any activity required by this Decree. In any action to enforce this Consent Decree, Settlors shall not raise as a defense the failure by any of its agents, servants, contractors, employees, successors or assigns to take actions necessary to comply with this Decree except as provided in Part XXI (Force Majeure).

IV. OBJECTIVES

7. It is the purpose of the parties in entering this Decree to further the objectives of RCRA, and the Clean Water Act and applicable State law. All plans, studies, construction, maintenance, monitoring programs, and other obligations in this Decree or resulting from the activities required by this Decree shall have the objective of causing Settlors to come into compliance expeditiously, and remain in full compliance with RCRA, the CWA, and other applicable State law.

V. DEFINITIONS

Unless otherwise defined herein, terms used in this Decree shall have the meaning given to those terms in RCRA the regulations promulgated thereunder, the Texas Solid Waste Disposal Act, the Texas Health and Safety Code Chapter 361, the Texas Water Code Chapter 26, or the CWA, and the regulations promulgated thereunder.

Acceptable shall mean that the quality of Submittals or completed work is sufficient to warrant review by the TNRCC and/or the EPA in order to determine whether the Submittal or work meets the terms and conditions of this Decree, including attachments, guidance documents, approved Work Plans and/or the TNRCC's or the EPA's written comments. Acceptability of Submittals or work, however, does not necessarily imply that will be approved. Approval by the TNRCC and/or the EPA, of Submittals or work, however, establishes that those Submittals were prepared, or work was completed, in a manner acceptable to the TNRCC and/or the EPA.

Administrative Record shall mean the record compiled and maintained by the EPA and the TNRCC in connection with an administrative decision regarding the implementation of this Decree.

AMS Standard shall mean, for purposes of this Decree only, the ASARCO Management System Standard dated November 17, 1997 attached to the Consent Decree entered in case number CV-98-3-H-CCL by the United States District Court for the District of Montana on May 5, 1998 and any future revisions thereto - which affect ASARCO's obligations under this Decree as identified and described more fully in Paragraphs 62-79 below. This shall not be construed to prevent ASARCO from modifying its AMS Standard for all other purposes.

Area of Concern (AOC) shall mean any area where a Release to the environment from the Facility of any Hazardous Waste or Hazardous Constituents has occurred, is suspected to have occurred, or may occur, regardless of the time, frequency or duration of the Release. The Areas of Concern include areas and/or discernible units at which solid wastes have been placed, at any time, irrespective of whether the area or unit was intended for the management of solid waste or Hazardous Waste. Examples of Areas of Concern include, but are not limited to, landfills, surface impoundments, pits, waste piles, land treatment units, incinerators, tank systems (including any storage, treatment, or accumulation tank system), container storage units, elementary neutralization units, waste or wastewater treatment systems that received solid waste or Hazardous Waste or Hazardous Constituents, or Released Hazardous Waste or Hazardous Constituents at any time.

ASARCO El Paso Facility shall mean the ASARCO, Incorporated, copper smelter, and contiguous areas under ASARCO ownership and control in El Paso, Texas.

ASARCO Operations shall mean unit(s) and complex(es) listed in Exhibit 8, which is incorporated by reference into this Consent Decree.

CAP shall mean the "RCRA Corrective Action Plan (Final)," OSWER Directive 9902.3-2A, dated May 1994.

Compliance Plan shall mean a corrective action plan which is issued to Hazardous Waste permitted facilities as required under Texas regulations 30 TAC §335.401 and implements compliance monitoring and corrective action requirements of 30 TAC §§335.156-167 (40 CFR §§264.90-101).

Corrective Measures shall mean those measures or actions appropriate to remediate, control, prevent, or mitigate the Release, potential Release or movement of Hazardous Waste or Hazardous Constituents into the environment or within or from one media to another.

EPA shall mean the United States Environmental Protection Agency, and any successor departments or agencies of the United States.

EMS Policy shall mean, for purposes of this Decree only, the Environmental Management System Policy dated December 12, 1997 attached to the Consent Decree entered in case number CV-98-3-H-CCL by the United States District Court for the District of Montana on May 5, 1998 (and any future revisions thereto - which affect ASARCO's obligations under this Decree as identified and described more fully in Paragraphs 62-79 below) . This shall not be construed to prevent ASARCO from modifying its EMS Policy for all other purposes.

Environmental Compliance Auditing Procedures shall mean, for purposes of this Decree only, the ASARCO Environmental Compliance Auditing Procedures, Exhibit 9 hereto (and any future revisions thereto - which affect ASARCO's obligations under this Decree as identified and described more fully in Paragraphs 62 to 79 below - upon mutual agreement of the parties and approval of the Court) which describes the corporate-wide environmental compliance audit program. This shall not be construed to prevent ASARCO from modifying its Environmental Compliance Audit Procedures for all other purposes.

Environmental Management System (EMS) shall mean the comprehensive corporate-level environmental management system required by and detailed in Paragraphs 152 through 174 of the Consent Decree entered in case number CV-98-3-H-CCL by the United States District Court for the District of Montana on May 5, 1998.

Environmental Requirements shall mean all applicable federal, state, and local environmental statutes and regulations, including permits and enforceable agreements between Settlors and the respective environmental regulatory agencies.

Facility as used in the term "the Facility" or in the term "the Encycle Facility" shall mean the Encycle/Texas, Incorporated, facility located at 5500 Up River Road in Corpus Christi, Texas.

Feedstock shall mean a material shipped to the Encycle Facility from suppliers, which is used for making Encycle Output, except for materials used at the Encycle Facility as effective substitutes for commercial products in accordance with 40 C.F.R. §261.2(e)(1)(ii); or materials legitimately used by Encycle's customers as flux substitutes and reagents in accordance with 40 C.F.R. §261.2(e)(1)(ii), as approved by the TNRCC, in consultation with EPA.

Ground Water shall mean the water in the saturated zone beneath the land surface.

Hazardous Constituents shall mean those constituents listed in Appendix VIII of 40 CFR Part 261 or any constituent identified in Appendix IX of 40 CFR Part 264 or any approved subset of Appendix IX of 40 CFR Part 264.

Hazardous Waste shall mean hazardous waste as defined in 40 CFR §261.3 and 30 TAC §335.1.

Imminent Threat shall mean any Release, or threatened Release, of solid waste or Hazardous Waste or a Hazardous Constituent, at or from the Facility, which may present an imminent or substantial endangerment to human health and/or the environment.

Interim Measure or IM shall mean those actions which can be, or are, initiated in advance of implementation of the final Corrective Measures.

Liquid Feedstocks shall mean Feedstocks that are not Solid Feedstocks.

Load shall mean a single discrete material received at the Encycle Facility on a single manifest, bill of lading or any other shipping document.

Output shall mean the managed or processed material at the Encycle Facility meeting customer specifications, which is shipped off-site for reclamation or use. Materials shipped off-site for treatment (not reclamation) and/or disposal are not Output.

Paragraph(s) shall mean a paragraph or paragraphs of this Decree numbered with arabic numerals, including all subparagraphs contained therein.

Part shall mean a portion of this Decree identified by an upper case roman numeral.

Pay Metal(s) shall mean chromium, cobalt, copper, lead, nickel, tin, and zinc, antimony, bismuth, selenium, tellurium, titanium, tungsten, vanadium, aluminum, molybdenum, cadmium and the Rare Earths as defined herein.

Permit shall mean the Hazardous Waste Storage Permit No. HW-50221-001 originally issued to Encycle, by the TNRCC on September 27, 1988 and subsequent modifications, renewals or amendments, approved by the TNRCC.

Precious Metal(s) shall mean gold, platinum, palladium, and silver, rhodium and ruthenium.

Rare Earths shall mean cerium group and yttrium group.

RCRA shall mean the Solid Waste Disposal Act, as amended by the Resource Conservation and Recovery Act, as further amended by, inter alia, the Hazardous and Solid Waste Amendments of 1984.

RCRA Facility Investigation or RFI shall mean the investigation and characterization of the source(s) and/or Releases of

Hazardous Wastes and/or Hazardous Constituents and the nature, extent, direction, rate, movement, and concentration of such Releases of Hazardous Wastes and/or Hazardous Constituents, that have been, or may be Released or may reasonably be expected to be Released into the environment.

RCRA metals include arsenic, barium, cadmium, chromium, lead, mercury, selenium, and silver.

Receptors shall mean those animals (including humans) or plants, and their habitats which are or may receive or be affected by Releases of Hazardous Waste or Hazardous Constituents.

Release shall mean any spilling, leaking, pumping, pouring, emitting, emptying, discharging, injecting, escaping, seeping, leaching, dumping, placing, or disposing into the environment.

Risk Assessment shall mean the process or framework for characterizing risk to human and ecological Receptors.

Settlors refers to Encycle and ASARCO.

Solid Feedstocks shall mean Feedstocks which contain minimal or no free liquids. Any free liquids, as determined pursuant to 40 CFR 260.11, present in arriving shipments of solid Feedstocks shall be absorbed (solidified) and/or decanted before the Feedstock may be placed as a Solid Feedstock in bins.

Submittal shall mean any Work Plan, report, progress report, or any other written document Settlors are required by this Decree to send to TNRCC and EPA.

TAC shall mean Texas Administrative Code.

Tennessee Mines Facilities shall mean the following mines and mills owned and operated by ASARCO in eastern Tennessee: Coy Mine, Young Mine, Young Mill, New Market Mine and Mill, and Young Mine (Beaver Creek Shaft), all located in Jefferson County, Tennessee; and Immel Mine, located in Knox County, Tennessee.

TNRCC shall mean the Texas Natural Resource Conservation Commission or the Executive Director for the Texas Natural Resource Conservation Commission as appropriate, and any successor departments or agencies.

Violation(s) shall mean actions, omissions, failures, or refusals to act by the Settlors that result in a failure to meet the terms and conditions of this Decree.

Work Plans shall mean the detailed plans prepared by the Settlors to satisfy requirements of this Decree. The requirements for each Work Plan are set forth in this Decree. All Work Plans and modifications or amendments thereto, shall be incorporated into this Decree by reference and are enforceable components of this Decree when approved in writing by the TNRCC and/or EPA, unless otherwise required under Part VII of this Decree.

[MATERIAL OMITTED]

XIV. AUDITING PROVISIONS

62. ASARCO shall conduct Environmental Management System (EMS) audits and Environmental Compliance audits at the Operations listed in Exhibit 8, as further specified below.

63. However, in the event ASARCO no longer has ownership interest or operational control at an Operation listed in Exhibit 8, ASARCO shall notify EPA in writing and provide supporting documentation of the transaction, including copies of the documents effecting the transfer. ASARCO, thereafter, shall no longer be obligated to perform the requirements contained in Part XIV of this Decree for that Operation.

64. Further, in the event that an Operation listed in Exhibit 8 has been placed on standby for a period of six (6) months or longer, or shut down, ASARCO shall notify EPA in writing and provide supporting documentation. ASARCO, thereafter, shall no longer be obligated to perform the requirements contained in this Part of this Decree for that Operation, unless ASARCO resumes operations during the pendency of this Decree at that Operation.

ENVIRONMENTAL MANAGEMENT SYSTEMS AUDITS

65. ASARCO shall develop and implement a five-year EMS audit program to assess whether an effective Environmental Management System (EMS) is being implemented at each ASARCO Operation listed in Exhibit 8. This program will commence effective January 1, 2000. The EMS audit program, as further described in the following Paragraphs, is structured into three phases with specific progressive objectives for each phase. The primary objective of the Phase 1 audits (first two audits of this program) shall be to provide ASARCO with feedback on EMS implementation progress, and with information needed for planning future improvements. ASARCO's goal is to have substantially implemented an EMS at its Operations listed in Exhibit 8 before the third audit year of this program. The primary objective of the Phase 2 and Phase 3 audits shall be: (a) to assess the EMS at each ASARCO Operation listed in Exhibit 8 to determine whether its implementation by ASARCO meets he criteria defined by the AMS Standard and ASARCO EMS Policy; (b) to identify any issues that may require further improvements or evaluations; and (c) to determine if the EMS is being maintained, and that continuous improvement has been institutionalized.

66. ASARCO shall ensure that the auditors performing the EMS audits and drafting the audit reports, required by this Consent Decree, are qualified, meaning that they satisfy the requirements of ISO 14012 (First edition, 1996-10-01) before conducting any of the audits required by this Decree, with the exception that on-the-job training shall be completed during the first three years after this Decree is entered by the Court. Classroom training shall include successful completion of an accredited ISO 14000 auditing course or equivalent within two (2) years after entry of this Decree by the Court. ASARCO shall

identify any and all site-specific training requirements for the auditors so they can effectively and safely conduct the required audits and shall ensure that their training is completed prior to conducting an audit.

67. ASARCO shall employ at least six (6) qualified auditors within the Operations listed in Exhibit 8 or the ASARCO Technical Services Center. These qualified auditors shall compose the group from which the audit teams shall be selected for conducting the company-led EMS audits required by this, Decree. Each audit team for Phase II audits shall have at least two (2) qualified auditors, one of which is designated and qualifies as a Lead Auditor, and neither of which is from the Operation being audited. Other qualified auditors or auditors in training from elsewhere in ASARCO may participate or assist in the audits.

68. As part of the audit program specified in Paragraph 65 above, ASARCO shall conduct or have conducted EMS audits at each ASARCO Operation listed in Exhibit 8 in accordance with subparagraphs A through C, below. General audit requirements are further identified in Paragraphs 69 and 70.

A. Phase 1 - Audit Years 1 and 2 (2000 and 2001). These EMS audits shall be conducted by qualified auditors, as described by Paragraphs 66 and 67, at each ASARCO Operation listed in Exhibit 8. These audits shall incorporate the following provisions and the objective shall be as stated in Paragraph 65.

1. To verify progress on those elements addressed by each Operation's EMS Development Plan (EMS Development Plans are required by Paragraphs 165 and 166 of the ASARCO Montana Consent Decree, Civil Action No. 98-3-H-CCL);

2. To evaluate, where systems have been implemented, conformance to the environmental criteria specified in the AMS Standard and ASARCO Environmental Management System Policy; and

3. To determine whether ASARCO is effectively communicating environmental requirements to affected parts of the organization, including contractors and on-site service providers.

B. Phase 2 - Audit Years 3, 4 and 5 (2002, 2003 and 2004). These annual EMS audits shall be conducted by qualified auditors, as described by Paragraphs 66 and 67, at each ASARCO Operation listed in Exhibit 8. These audits shall incorporate the following provisions and shall be conducted in accordance with ISO 14011 (First edition, 1996-10-01), using ISO 14010 (First edition, 1996-10-01) as supplemental guidance. Each audit's objectives shall be as stated in Paragraph 65.

The EMS audit shall encompass an evaluation of the adequacy of EMS implementation, from the Operations manager down through the Operation. The auditors shall determine conformance of the EMS with the AMS Standard and ASARCO EMS Policy, by assessing the following:

1. Whether there is a defined program or planned task for the EMS element;
2. Whether the program or task has been implemented, is being maintained, and to what extent;
3. Adequacy of each Operation's internal self-assessment procedures for programs and tasks composing the EMS;
4. Whether ASARCO is effectively communicating environmental requirements to affected parts of the organization, including contractors and on-site service providers;
5. Whether further improvements should be made to the EMS; and,
6. Whether there are observed deviations from ASARCO's written requirements or procedures.

C. Phase 3 - Audit Years 4 and 5 (2003 and 2004). By the end of Phase 3, one EMS audit shall be conducted at each Operation listed in Exhibit 8. ASARCO shall contract for a qualified, independent, third-party auditor(s) (Contract Auditor) to conduct these audits. The Contract Auditor will be selected and paid by ASARCO. ASARCO shall provide a copy of this Consent Decree to each Contract Auditor who is retained to conduct any of the EMS audits required by this Consent Decree. Not later than 30 days before the first audit conducted by the Contract Auditor, ASARCO shall provide to EPA in writing: (1) the name, affiliation and address of the Contract Auditor(s) selected by ASARCO to conduct the Phase 3 audits at each Operation listed in Exhibit B and (2) demonstrate how the selected Contract Auditor satisfies the qualification requirements of ISO 14012 (First edition, 1996-10-01). Each Contract Auditor must be capable of exercising the same independent judgment and discipline that a certified public accounting firm would be expected to exercise in auditing a publicly held corporation.

The audits conducted by the Contract Auditor shall conform to the requirements stated in Paragraph 68 B above, and shall not be conducted concurrently with the Phase 2 audits. The Contract Auditors may also report on areas of concern identified during the course of the audits, that, in their judgement [sic], merit further review or evaluation for potential environmental or regulatory impacts.

69. ASARCO shall prepare and provide to EPA, by January 3 of each year of the Phase 2 and 3 audit programs required by this Consent Decree, a schedule indicating the month during which the EMS audit will be conducted at each ASARCO Operation. The schedule may be revised by ASARCO, provided the required EMS audits are conducted in accordance with this Consent Decree.

70. For each EMS audit in Phase 2 and Phase 3, ASARCO shall prepare, or have prepared, a written EMS Audit Plan, Audit Report, and EMS Audit Action Plan for the Operation. Each Action Plan shall be developed in a timely fashion and corrective actions shall be implemented in an expeditious manner.

71. ASARCO shall prepare an Annual Report for the Phase 2 and 3 EMS audits required by this Consent Decree. The Annual Report shall be submitted to EPA by January 31 of the year following the year in which the audits were conducted (i.e., reports are due in 2003, 2004 and 2005). The report shall list the dates and locations at which audits were conducted, the personnel who conducted each audit, and the date the Action Plan was drafted. It shall also contain a certification by a responsible corporate officer that all required audits were conducted in accordance with the provisions of this Decree.

COMPLIANCE AUDITS

72. ASARCO shall conduct a five-year compliance audit program to assess ASARCO's compliance at all ASARCO Operations listed in Exhibit 8. This program will commence effective January 1, 2000. ASARCO shall follow the Environmental Compliance Auditing Procedures, attached hereto as Exhibit 9, when conducting these compliance audits at ASARCO Operations.

73. ASARCO shall ensure that the auditors performing the environmental compliance audits and drafting the audit reports, required by this Consent Decree, are qualified, meaning that they satisfy the requirements of ISO 14012 (First edition, 1996-10-01) before conducting any of the audits required by this Decree, with the exception of the on-the-job training requirements. All lead auditors shall satisfy the on-the-job training requirements before conducting any of the audits required by this Decree. Classroom training may include internally developed instruction on environmental compliance auditing. ASARCO shall identify any and all site-specific training requirements for the auditors so they can effectively and safely conduct the required audits and shall ensure that their training is completed prior to conducting the audit.

74. ASARCO shall employ at least six (6) qualified auditors within the Operations listed in Exhibit 8 or the ASARCO Technical Services Center. These qualified auditors shall compose the group from which the audit team shall be selected for conducting the company-led environmental compliance audits required by this Decree. Each audit team shall have at least two (2) qualified auditors, one of which is designated and qualifies as a Lead Auditor, and neither of which is from the Operatic. being audited. Other qualified auditors or auditors in training from elsewhere in ASARCO may participate or assist in the audits.

75. ASARCO shall prepare and provide to EPA an annual schedule indicating the month during which the compliance audit will be conducted at each ASARCQ Operation, by January 31 of each year, beginning in 2000 and ending in 2004. The schedule may be revised by ASARCO, provided the required compliance audits are conducted in accordance with this Consent Decree.

The general audit schedule shall be as follows, for Operations listed in Exhibit 8.

YEAR	OPERATIONS AUDITED
2000	Group 1
2001	Groups 1 and 2
2002	Group 1
2003	Groups 1 and 2
2004	Group 1

76. ASARCO shall develop Audit Reports and Action Plans for each environmental compliance audit required by this Decree in accordance with the provisions of the Environmental Compliance Auditing Procedures. Each Action Plan shall be developed in a timely fashion and corrective actions shall be implemented in an expeditious manner.

77. ASARCO shall submit annual reports to EPA containing a certification by a responsible corporate officer that the compliance audits conducted during the reporting period pursuant to this Decree, were performed in accordance with the Environmental Compliance Auditing Procedures and in accordance with the provisions of this Decree for all ASARCO Operations listed in Exhibit 8 audited during the reporting year. The Annual Reports shall be submitted to EPA by January 31 each year, beginning in 2001 and ending in 2005. The report shall list the dates and locations at which audits were conducted, the personnel who conducted each audit, and the date the Action Plan was written.

78. ASARCO shall enter into a contract with a qualified, independent, third-party audit consultant ("Consultant Auditor") to conduct all Compliance Audit Program Conformance Reviews of ASARCO's five (5) environmental compliance audits required by this Decree. The objective of these Conformance Reviews shall be to determine if ASARCO's compliance audits are being conducted in conformance with the Environmental Compliance Auditing Procedures and the following provisions.

A. Compliance Audit Program Conformance Review Schedule - The Conformance Reviews shall be conducted in accordance with the following schedule.

		CONFORMANCE REVIEWS*	
	COMPLIANCE AUDITS	GROUP 1	GROUP 2
2001	Groups 1 and 2	2	2
2002	Group 1	2	0
2003	Groups 1 and 2	2	1
2004	Group 1	1	0

For Group 1 Operations, a Conformance Review of the environmental compliance audits at the East Helena Lead Smelter and the Encyle/Texas Plant shall be conducted twice during the five-(5-)year compliance audit program required by this Decree. For Group 2 Operations, one of the Conformance Reviews shall be conducted on a compliance audit of the Tennessee Mines Division conducted during the compliance audit program required by this Decree. The Consultant Auditor shall select other Operations whose compliance audits will be reviewed for conformance with the Environmental Compliance Auditing Procedures.

B. Selection of Consultant Auditor

ASARCO, prior to September 1, 2000 , shall propose to EPA for approval, the selection of an independent Consultant Auditor who meets the qualification requirements of ISO 14012 (First edition, 1996-10-01); and has expertise and competence in the regulatory programs under federal and state environmental laws. The Consultant Auditor will be paid by ASARCO. The Consultant Auditor must not directly own any stock in ASARCO or in any parent or subsidiary, and must have no other direct financial stake in the outcome of the compliance audits conducted by ASARCO pursuant to this Consent Decree. The Consultant Auditor must be capable of exercising the same independent judgment and discipline that a certified public accounting firm would be expected to exercise in auditing a publicly held corporation. If ASARCO has any other contractual relationship with the Consultant Auditor, ASARCO shall disclose to EPA such past or existing contractual relationships. EPA will notify ASARCO in writing of its approval or disapproval as expeditiously as possible. If, in EPA's judgment, such past or existing relationship would affect the Consultant Auditor's ability to exercise the independent judgment and discipline required to conduct the Compliance Audit Program Conformance Review, such Consultant Auditor shall not be retained by ASARCO and another Consultant shall be proposed within thirty (30) days of ASARCO's receipt of EPA's determination. If ASARCO wishes to contract a new Consultant Auditor, ASARCO shall notify EPA in writing and provide an explanation for the change, and shall propose another Consultant Auditor to EPA for approval. Any subsequent Consultant Auditor must satisfy the qualification requirements of this Paragraph.

C. Conformance Review Workplan

ASARCO shall provide a copy of this Consent Decree to each Consultant Auditor who is retained to conduct any of the conformance reviews required by this Consent Decree. Within sixty (60) days prior to the beginning of the first Conformance Review, the Consultant Auditor shall prepare and submit to ASARCO and EPA, for review and comment, a proposed Workplan for the Compliance Audit Program Conformance Review. The proposed Workplan shall specify the Consultant Auditor's plan for conducting the conformance reviews, and shall be designed to meet the objectives and criteria set forth in this Part. The Workplan shall describe the procedure for annually selecting the Operations whose compliance audits will be reviewed for conformance with the Environmental Compliance Auditing Procedures. The Consultant Auditor shall follow the approved workplan when selecting Operations' audits and conducting the Conformance Reviews required by this Decree. ASARCO shall cooperate with the Consultant Auditor in connection with the preparation and submission of this Workplan, and its subsequent execution. ASARCO shall provide the Consultant Auditor documents necessary for conducting the conformance review, including, but not limited to, audit procedures, reports, and action plans, as well as access to audited Operations and staff participating in the compliance audits.

D. Compliance Audit Program Conformance Review Report Submittal

The Consultant Auditor shall prepare and submit a Conformance Review Report to ASARCO in a timely fashion. ASARCO, thereupon, shall submit the Conformance Review Report prepared by the Consultant Auditor to EPA by January 31 each year, beginning in 2002 and ending in 2005. This report shall describe ASARCO's conformance and instances of nonconformance with the Environmental Compliance Auditing Procedures, and shall contain the following information listed in ISO 14010 (First Edition, 1996-10-01), Section 4.7 ("Audit Report")

1. The identification of the Operations reviewed by the Consultant Auditor;
2. The objectives and scope of the Conformance Review;
3. The criteria against which the Conformance Review was conducted;
4. The period covered by the Conformance Review and the date(s) it was conducted;
5. The identification of the Conformance Review team members;
6. The identification of the ASARCO representatives participating in the Conformance review;
7. A statement of the confidential nature of the contents;
8. The distribution for the Conformance Review Report;
9. A summary of the Conformance Review process, including any obstacles encountered;
10. The conclusions of the Conformance Review, including the conformance and instances of non-conformance, if any, with the Environmental Compliance Auditing Procedures; and
11. The timeliness and thoroughness of the compliance audits, whether an Action Plan was developed and implemented, and whether the Action Plan was effective when completed.

[MATERIAL OMITTED]

XXIX. CONTINUING JURISDICTION OF THE COURT

141. The Court shall retain jurisdiction to enforce this Decree and to resolve disputes arising hereunder as may be necessary or appropriate for the construction or execution of this Decree.

XXX. TERMINATION

142. This Decree shall be terminated in two stages as set forth in this Part upon completion of all requirements of this Decree applicable to each stage as follows:

Stage I. As to the requirements of this Decree regarding compliance by Encycle's operations with the hazardous waste management standards of RCRA and State law as set forth in Sections VI and Section X, this Decree shall terminate upon joint motion of the parties any time after five years from the date of entry of this decree provided:

a. TNRCC or EPA has issued Encycle a new hazardous waste management permit under RCRA covering Encycle's operations at the Corpus Christi Facility;

b. Settlors are in compliance with the terms of this permit and the terms of this Decree;

c. Settlors have submitted a certification to the United States and the States that conditions a. and b. above have been met; and

d. the United States and State, has concurred in writing with Settlors' contention that conditions a. and b. have been met. If the United States or the State, disputes Settlors' contention, the dispute resolution provision of this Decree (Part XXII) shall be invoked and this Decree shall remain in effect pending resolution of the dispute by the parties or the Court.

Stage II. As to all other provisions of this Decree, upon joint motion of the parties after each of the following has occurred:

a. Settlors have completed all actions required by this Decree;

b. Settlors have paid all monies and penalties due under this Decree;

c. Settlors have submitted a certification to the United States and the State that conditions a. and b. above have been met; and

d. The United States and State, has concurred in writing with Settlors' contention that conditions a. and b. have been met. If the United States or the State, disputes Settlors' contention, the dispute resolution provision of this Decree (Part XXII) shall be invoked and this Decree shall remain in effect pending resolution of the dispute by the parties or the Court.

Dated and entered this 6th day of October

UNITED STATES DISTRICT JUDGE

[MATERIAL OMITTED]

EXHIBIT 9

OBJECTIVES AND SCOPE OF THE ENVIRONMENTAL COMPLIANCE AUDIT PROGRAM
Document Number ENV-800-G-PY-001

1. Purpose
This document defines the objectives and scope of Asarco's environmental compliance audit program.

2. Scope
This policy applies to Asarco's internal environmental compliance audit program, managed by a Technical Services Center project manager at the direction of the Legal Department.

3. Definitions
3 .1 Environmental Compliance Audit: An environmental compliance audit is an objective, systematic examination of unit facilities, processes, procedures, and practices to verify whether they comply with legal requirements, internal policies, and accepted practices.
3.2 Audit Finding: An audit finding is primarily an exception-based, written summary of one or more non-conformances noted during an environmental compliance audit. A finding may also include an area of concern identified during the course of the audit that, in the auditor's judgement [sic], merits further review or evaluation for potential environmental or regulatory impacts
3.3 Applicable Regulatory Requirements: Applicable regulatory requirements are existing, new, and changing federal, state, and local environmental statutes and regulations, including permits and enforceable agreements between Asarco and respective environmental regulatory agencies.
3.4 Accepted Best Management Practice (BMP): A best management practice is a method of handling a potential environmental risk that incorporates a policy, procedure, equipment, and or personnel to go beyond compliance—either by managing a risk in a manner that exceeds the applicable regulatory requirements or by managing a risk not subject to regulatory requirements. To be an *accepted* BMP, the method must be effectively employed by another facility within the industry or commonly employed by facilities outside the industry.

4. Responsibility
4.1 Environmental compliance audits are to be conducted in accordance with ENV-800-G-PR-001, using established ASARCO Environmental Compliance Audit Protocols.

5. References/Appendices
5.1 Environmental Safety and Health Policy
5.2 ENV-400-G-PY-002: Environmental Management System Policy
5.3 ENV-800-G-PR-001 Environmental Compliance Auditing
5.4 ASARCO Environmental Compliance Audit Protocols

6. Policy
6.1 In support of the objectives outlined in Asarco's Environmental, Safety and Health Policy and the requirements of Asarco's Environmental Management System Policy, the objectives of the environmental compliance audit program are:
6.1.1 To verify independently each unit's compliance with applicable regulatory requirements, company policies, and accepted best management practices on a periodic basis.
6.1.2 To report audit findings expeditiously to the appropriate managers and, if required or voluntarily disclosed, to the appropriate regulatory agency(ies).
6.1.3 To resolve every audit finding in a timely fashion by correcting its root cause.
6.1.4 To expand unit environmental staff's knowledge of environmental requirements and environmental management practices through participation in audits of other units.
6.1.5 To spread environmental knowledge and good environmental management practices from one unit to another.
6.2 The scope of the audits includes the following regulatory areas, where applicable:
Air
Air Pollution Control
Asbestos Management (complementary to safety and health audits)
Water
Water Pollution Control (including Stormwater Management)
Drinking Water Management
Waste
Solid and Hazardous Waste Management
Underground Injection Control
Storage Tanks
Spill Prevention and Control (including Aboveground Storage Tanks)
Underground Storage Tanks
DOT
Hazardous Materials Transportation
TSCA
TSCA Management
PCB Management
CERCLA/EPCRA
(including SARA Sections 302 and 303, Tier HI Reporting, and Form R/TRI
Toxic Chemical Release Reporting)
FIFRA
Other Applicable Federal Requirements
Specific State/Local Requirements
Process Evaluations

To verify identification of wastestreams and emission points (review process flow diagrams and regulatory interpretations)

7. Records/Reports
None

8. Revision History/Reason for Change
8.1 This is the first version of this document.

ENVIRONMENTAL COMPLIANCE AUDITING
Document Number: ENV-800-PR-001

1. Purpose
This document defines the procedure for conducting environmental compliance audits.

2. Scope
This procedure applies to audits conducted as part of Asarco's internal environmental compliance audit program, managed by a Technical Services Center project manager at the direction of the Legal Department. This procedure is intended to provide appropriate quality assurance and quality control (QA/QC) for this program by ensuring that the environmental compliance audits are properly staffed, planned, and conducted in accordance with ASARCO Environmental Compliance Audit Protocols, that draft audit reports are thoroughly reviewed, appropriate action plans are expeditiously developed and implemented, and that audit quality is high and consistent. Supplemental QA/QC activities not specifically addressed here include management Systems audits; team lawyer annual reviews; and continuing oversight by the division environmental oversight coordinators, division environmental managers, and the unit environmental managers, for which they are responsible and accountable.

3. Definitions
3.1 None

4. Responsibility
4.1 The assigned Technical Services Center project manager is responsible for scheduling audits and selecting qualified auditors independent of the unit being audited in accordance with ENV-800-G-PR-002, pre-audit preparation in accordance with ENV-800-G-PR-003, and giving guidance and support to the audit teams as needed.
4.2 The audit team leader is responsible for serving as the main team contact, pre-audit preparation in accordance with ENV-800-G-PR-003, leading the on-site opening and closing meetings, participating as an auditor, giving guidance and support to the rest of the audit team as needed, collecting the audit notes, and writing and issuing the audit report at the direction of the audit team lawyer in accordance with ENV-800-G-PR-004.
4.3 The audit team lawyer is responsible for pre-audit preparation in accordance with ENV-800-G-PR-003, participating as an auditor, assisting the team in interpretation of legal issues, consulting with outside counsel as necessary, and advising the audit team leader in writing and issuing the audit report in accordance with ENV-800-G-PR-004.
4.4 For each audit in the division he/she oversees, when not acting as the audit team leader, the environmental oversight coordinator is responsible for acting as an audit team advisor: assisting the project manager with scheduling audits and selecting auditors in accordance with ENV-800-G-PR-002, assisting the audit team leader with pre-audit preparation in accordance with ENV-800-G-PR-003, participating as an auditor and reviewing the unit's action plan and implementation progress in accordance with ENV-800-G-PR-005.
4.5 Each auditor is responsible for pre-audit preparation in accordance with ENV-800-G-PR-003 and participation in the audit.
4.6 The unit manager is responsible for pre-audit preparation in accordance with ENV-800-G-PR-003, facilitating the site visit, and drafting and implementing an action plan in accordance with ENV-800-G-PR-005.
4.7 For each audit of a unit he/she oversees, the unit manager's immediate supervisor is responsible for approving the unit's action plan and for reviewing the unit's implementation progress in accordance with ENV-800-G-PR-005.
4.8 For each audit of a unit in the division he/she oversees, the division environmental manager is responsible for reviewing the draft audit report in accordance with ENV-800-G-PR-004 and for reviewing the unit's progress in implementing its action plan in accordance with ENV-800-G-PR-005.

5. References/Appendices
References 5.1 and 5.2 are format examples subject to changes and are not included in this Decree.
5.3 ASARCO Environmental Compliance Audit Protocols
5.4 ENV-800-G-PR-001: Objectives and Scope of the Environmental Compliance Audit Program
5.5 ENV-800-G-PR-002: Environmental Compliance Audit Scheduling and Auditor Selection
5.6 ENV-800-G-PR-003: Pre-Audit Preparation for Environmental Compliance Audits
5.7 ENV-800-G-PR-004: Environmental Compliance Audit Report Contents, Distribution, and Document Control
5.8 ENV-800-G-PR-005: Environmental Compliance Audit Follow-up/Action Plans
5.9 Little, A. D., 1996, *EHS Auditing: Fundamentals, Skills, arid Techniques*

6. Procedure/Work Instruction
6.1 Each unit is reviewed for compliance in the applicable regulatory areas (see ENV-800-G-PR-001) at least once every two years, either via one full-scope environmental compliance audit or multiple partial-scope environmental compliance audits.
6.2 On an on-going basis, the Technical Services Center project manager, at the direction of the legal department, works with the units and audit teams to schedule and plan environmental compliance audits.
6.2.1 Each audit team is comprised of a team leader, team lawyer, the appropriate division environmental oversight

coordinator, when not acting as the audit team leader, and at least one more auditor, selected in accordance with ENV-800-G-PR-002.
6.2.2 Pre-audit preparation is done in accordance with ENV-800-G-PR-003.

6.3 Opening Meeting
The audit begins with an on-site opening meeting, chaired by the audit team leader, attended by the audit team, unit manager, and other appropriate personnel.
6.3.1 The unit provides the audit team with an overview of operations and environmental issues.
6.3.2 The audit team leader outlines the audit process. The following items are typically addressed:
· Purpose and scope of the audit
· Team member assignments
· General agenda of the audit
· Arrangements for staff interviews, as necessary
· Document control
· Arrangements for the closing meeting (time, place, teleconferencing of senior management, etc.)

6.4 After the opening meeting, the team leader confirms arrangements for safety equipment, safety training (if necessary), and work space for the team.

6.5 Initial Tour
The unit provides the audit team with a site plot plan and an initial tour of the facilities. The auditors gain a general orientation to the unit, including its layout and size, location of operations, location of activities pertaining to the audit scope, and location of employees whom may need to be interviewed. For guidance, the auditors should take along the site plot plan and "Table 2; Tour Checklist" from the ASARCO Environmental Compliance Audit Protocols.
6.5.1 If all facilities within the unit are close, the team uses the initial tour for obtaining an overview only. The auditors keep track of areas they will need to revisit individually later in the audit.
6.5.2 If the unit is spread out, the initial tour may serve as the main inspection. In this case, the auditors take detailed notes on remote areas to minimize the number of times they will need to revisit those areas of the unit.

6.6 Understanding Management Systems
Using "Step I: Understand Management Systems" from the ASARCO Environmental Compliance Audit Protocols as a guide, the auditors develop a detailed understanding of the unit's overall approach to managing its regulatory obligations. This helps the auditors to:
· Prioritize the audit topics they are to review
· Determine who the key personnel are for each environmental topic
· Gain insights regarding how effectively each environmental topic is being managed
· Establish a context for evaluating audit results
· Identify potential underlying causes that contribute to compliance-related deficiencies
The specific objectives of this step are to understand:
· How the unit has evaluated the applicability of existing regulatory standards to unit operations
· Types of unit procedures that exist to implement compliance activities. Whether compliance management roles and responsibilities are clearly understood
· Types of programs or activities used to familiarize staff with environmental compliance information
· How the unit maintains and documents the effectiveness of their environmental compliance activities

6.7 Gathering Audit Evidence
Once the auditors feel that they have a good understanding of the unit's management system, usually by the end of the first day or beginning of the second day, they begin to gather audit evidence. This step is the main focus of the audit and typically takes more time to accomplish. Using "Step 2: Gather Audit Evidence" from the ASARCO Audit Protocol(s) as a guide, the auditors evaluate and prioritize what needs to be done and begin gathering evidence by reviewing records, interviewing employees, and revisiting areas of the unit, as necessary. It is crucial that the auditors keep good notes on what they find, source(s) of information, and sampling strategies used in gathering evidence.

6.8 Evaluating Audit Findings
As the auditors complete gathering audit evidence for a topic or a group of topics, each one reviews the actions taken to ensure that sufficient and appropriate data have been gathered, working independently on his/her assigned topics and using "Step 3 Evaluate Audit Findings" from the ASARCO Audit Protocol(s) as a guide.
6.8. The auditors first make sure they have correctly identified the regulatory and internal requirements relevant to the particular topic(s).
6.8.2 They summarize their conclusions regarding the facility's compliance status for the topic(s).
6.8.3 They make sure they have collected an appropriate mix of relevant physical, testimonial, and documentary evidence. In order to be sufficient to support an audit finding, audit evidence must be relevant, objective, and persuasive.
6.8.4 They review and discuss any unresolved compliance issues with appropriate facility personnel.

6.8.5 They develop written lists of their exceptions and observations to use as a basis for discussion with other team members prior to the closing meeting.

6.9 Summarizing Findings as a Team

The audit team develops a comprehensive, written list of all audit findings and types of supporting evidence for presentation and discussion at the closing meeting. If the auditors evaluating their results individually all finish at the same time, the team should sit down as a group to discuss and summarize their findings. An alternate approach is for the team leader and team lawyer to summarize findings and types of supporting evidence with each auditor, as each one is finished.

6.9.1 The team critically reviews the auditors' conclusions to confirm that all findings are substantiated by the evidence collected.

6.9.2 The team prepares written findings to present to the unit at the closing meeting. Because the written findings provide the basis for the closing meeting discussion, it is imperative that the team is careful to write each finding clearly and concisely and to make sure that the findings are factually accurate and adequately supported by evidence. At the discretion of the team leader, some issues may be held to present verbally, such as issues noted that are isolated anomalies, outside of the audit scope, or very minor.

6.9.3 In reviewing the findings, the team must look for situations where multiple findings listed relate to one basic problem, possibly representing a system deficiency. At the discretion of the team leader, such findings may be grouped into one finding that points to a more common problem. For example, the audit team may have several findings all related to training, which point to one underlying management systems deficiency: that the unit has insufficient means to identify and track environmental training.

6.10 Prior to the closing meeting, the audit team should review the written findings with the unit employee(s) responsible for environmental compliance. He/she will want to know what the team plans to present to the unit manager before it happens. Also, he/she might have questions/comments on the findings that the team needs to clarify prior to the closing meeting.

6.11 Closing Meeting

At the end of the audit, the audit team meets with unit manager and other appropriate personnel to share all of their findings and observations. This gives the unit a preview of the audit report and allows for a final, pre-report confirmation that each of the findings is legitimate and factual.

6.11.1 To open the meeting, the team leader should:

- **Thank facility personnel for their help, support, and cooperation**
- **Briefly summarize the audit goals, scope, and approach**
- **Clarify that this meeting is the unit's opportunity to hear all of the findings and to understand and clarify any facts surrounding the details of the findings**

6.11.2 The findings may be presented one topic at a time by the auditor assigned to that topic or by the team leader. Each finding and summary of supporting evidence should be presented succinctly, followed by discussion and clarification, as needed, until everyone understands the nature of the finding. For each topic, the team should also briefly outline any positive items, such as programs that were recognized as strong or exceeding the requirements.

6.11.3 At the end of the meeting, the team leader collects all notes and computer files generated by the audit team.

6.12 Reporting Audit Findings

After the audit, the team leader generates a written audit report at the direction of the audit team lawyer in accordance with ENV-800-G-PR-004.

6.13 Follow-Up/Action Plans

In accordance with ENV-800-G-PR-005, the unit staff conducts a root cause analysis of the findings presented in the audit report, then develops an action plan, the implementation of which is checked regularly by the unit manager, unit environmental coordinator and the division environmental manager.

7. Records/Reports

7.1 ASARCO Environmental Compliance Audit Protocols. Master copies of these are maintained and distributed by the project manager.

7.2 Audit Notes/Working Papers. The team leader collects and files all notes and computer files generated during the audit by the audit team.

8. Revision History/Reason for Change

8.1 This is the first version of this document.

ENVIRONMENTAL COMPLIANCE
AUDIT SCHEDULING AND AUDITOR SELECTION
Document Number: ENV-800-G-PR-002

1. Purpose
This document defines the procedure for scheduling environmental compliance audits and auditor selection.

2. Scope
This procedure applies to audits conducted as part of Asarco's internal environmental compliance audit program, managed by a Technical Services Center project manger at the direction of the Legal Department.

3. Definitions
3.1 None

4. Responsibility
4.1 Under the direction of the Legal Department, the project manager is responsible for annually: establishing the list of audits to be held; setting the scope of each; ordering the audits by month; assigning audit team members; designating the audit team leaders; creating, issuing, and maintaining the audit schedule; establishing audit dates; and announcing each audit via a letter.
4.2 The environmental oversight coordinators advise the project manager in scheduling the audits, assigning audit team members, and establishing audit dates.
4.3 The unit advises the project manager in setting the audit dates.

5. References/Appendices
Reference 5.1 is a format example subject to change and is not included in this Decree.
5.2 ENV-800-G-PR-001: Environmental Compliance Auditing
5.3 ISO 14012 (First edition, 1996-10-01)

6. Procedure/Work Instruction
6.1 Under the direction of the Legal Department, in October of each year the Technical Services Center project manager establishes the list of audits to be held the following year and sets the scope of each. The schedule is submitted to the ASARCO Environmental, Safety and Health Policy Review Committee (as appointed by the ASARCO Management Committee) for review and approval.
6.2 Once the schedule has been approved, the project manager (a qualified auditor, as described in 6.3, below) and the environmental oversight coordinators schedule the audits by month and assign audit team leaders who satisfy the requirements of ISO 14012 (First edition, 1996-10-01). The division environmental oversight coordinator may serve as the audit team leader.
6.3 The project manager assigns at least two qualified audit team members to each team, including one who qualifies as a team leader, taking into consideration qualifications and expertise of certain individuals and all requests made by the Legal Department, the environmental oversight coordinators, and units. "Qualified" means that they satisfy the requirements of ISO 14012 (First edition, 1996-10-01), with the exception of the on-the-job training requirements. Other qualified auditors or auditors in training from elsewhere in ASARCO may participate or assist in the audits.
6.4 The project manager issues the audit schedule to all units being audited and/or personnel assigned as audit team members. The schedule is updated and redistributed as necessary by the project manager.
6.5 Prior to each audit (usually 1-3 months), the project manager contacts the unit to be audited along with each key audit team member and determines the optimal dates for the audit.
6.6 The project manager then sends a letter to the unit manager to confirm the audit dates, identify the audit team leader and members, and describe the audit's agenda, scope, and initial plans. The letter is copied to the environmental operations vice president, division environmental manager, audit team lawyer, lawyer responsible for directing the audit program, audit team members, and others, if necessary.

7. Records/Reports
7.1 Audit Schedule. This is maintained and filed by the project manager.
7.2 Audit Confirmation Letter. These are issued and filed by the project manager.

8. Revision History/Reason for Change
8.1 This is the first version of this document.

PRE-AUDIT PREPARATION
FOR ENVIRONMENTAL COMPLIANCE AUDITS
Document Number: ENV-800-G-PR-003

1. Purpose
This document defines the procedure for pre-audit preparation for environmental compliance audits.

2. Scope
This procedure applies to audits conducted as part of Asarco's internal environmental compliance audit program, managed by a Technical Services Center project manager at the direction of the Legal Department.

3. Definitions
3.1 None

4. Responsibility
4.1 The Technical Services Center project manager is responsible for confirming the availability of audit team members, discussing the audit plans with the audit team leader, and giving guidance and support to the audit team as needed.
4.2 The audit team leader is responsible for taking care of logistical arrangements and establishing the audit agenda and team member assignments.
4.3 The audit team lawyer is responsible for reviewing the most recent applicable audit report(s) and confirming regulatory applicability with local counsel, as necessary. This information is shared with audit team members.
4.4 When not acting as the audit team leader, the division environmental oversight coordinator is responsible for providing the audit team leader with environmental information about the unit and advising the project manager and audit team leader as needed.
4.5 Each audit team member is responsible for reviewing available unit materials pertinent to his/her assigned area(s), along with the appropriate ASARCO Environmental Compliance Audit Protocol(s). This will necessarily include applicable regulatory requirements, unit processes, and waste streams and emission points.
4.6 The unit to be audited is responsible for providing the audit team leader and division environmental oversight coordinator with information, as needed, and for facilitating the audit visit. The unit is responsible for ensuring its environmental files will be organized and accessible, providing the audit team with its own workspace, and making appropriate unit personnel available for interviews during the audit. The unit is responsible for planning to operate under normal conditions during the audit.

5. References/Appendices
Reference 5.1 is a format example subject to change and is not included in this Decree.
5.2 ENV-800-G-PR-001: Environmental Compliance Auditing
5.3 ASARCO Environmental Compliance Audit Protocols

6. Procedure/Work Instruction
6.1 Prior to each audit, the Technical Services Center project manager confirms the availability of audit team members, discusses the audit with the audit team leader, sends the audit team leader the names and phone numbers of the audit team members and unit personnel involved, and gives guidance and support to the audit team as needed.
6.2 The audit team leader serves as the main team contact. He/she is in charge of pre-audit preparation, including:
· Making/confirming necessary arrangements for team logistics (such as car rental(s), hotel reservations, maps, meeting times, etc.).
· Confirming arrangements for the opening meeting (time and place) with the unit manager.
· Making arrangements for safety briefing and security clearances with the unit environmental manager.
· Reviewing available permits and unit environmental issues with input from division environmental oversight coordinator.
· Assigning audit team members to regulatory areas in consultation with division environmental oversight coordinator.
· Providing the appropriate ASARCO Environmental Compliance Audit Protocol(s) to the audit team members, in accordance with their assignments.
· Preparing and sending an audit agenda, list of required safety equipment, regulatory assignments, etc. (see Appendix A) to the audit team members.
6.3 The audit team reviews copies of the most recent audit report and action plan for the unit, provided by the audit team lawyer, and, if necessary, previous reports.
6.4 The audit team lawyer, in consultation with the audit team leader or appropriate team member, confirms that determinations by the unit environmental manager regarding applicability of the state and local requirements are correct.
6.5 Reserved
6.6 Each audit team member reviews available unit materials pertinent to his/her assigned area(s), along with the ap-

propriate ASARCO Environmental Compliance Audit Protocol(s).
6.7 The unit to be audited provides the audit team leader with orientation materials, such as an organization chart, facility map, and process flow diagrams prior to the audit. (If this is infeasible, the orientation materials may be distributed to the audit team during the opening meeting.)
6.8 The unit makes plans to ensure that its environmental files will be organized and accessible during the audit, including:

- **Environmental permits**
- **Written policies relating to environmental procedures, emergency plans, etc.**
- **Written management systems, SOPs, or task lists for environmental compliance**
- **Recent environmental surveys, reports, inspections, or compliance checklists**
- **Notices of violation, compliance orders, notices of abatement, etc.**

Previous action plans (including root cause analysis, as appropriate) addressing findings from previous audits.

- **Training records**

6.9 The unit makes plans to facilitate the audit by providing the audit team with its own workspace and by making appropriate unit personnel available for interviews during the audit.
6.10 The unit should be operating under normal conditions during the audit. If otherwise. the unit manager should request rescheduling of the audit, in accordance with ENV-800-G-PY-003 Environmental Compliance Audit Cancellation Policy.

7. Records/Reports
7.1 Pre-Audit Memo to Audit Team. If the audit team leader issues a pre-audit memo, a copy is filed by the project manager.

8. Revision History/Reason for Change
8.1 This is the first version of this document.

ENVIRONMENTAL COMPLIANCE AUDIT REPORT CONTENTS, DISTRIBUTION, AND DOCUMENT CONTROL

Document Number: ENV-800-G-PR-004

1. Purpose
This document defines the procedure for environmental compliance audit report contents, distribution, and document control.

2. Scope
This procedure applies to audit reports generated as the result of audits conducted as part of Asarco's internal environmental compliance audit program, managed by a Technical Services Center project manager at the direction of the Legal Department.

3. Definitions
3.1 None

4. Responsibility
4.1 At the direction of the audit team lawyer, the audit team leader is primarily responsible for writing and issuing the audit report and for collecting all audit-related documents and evidence.
4.2 The audit team lawyer is responsible for advising the audit team leader in writing and issuing the audit report.
4.3 The audit team members are responsible for assisting the team leader in developing and reviewing the draft audit report, as needed.
4.4 The division environmental manager responsible for reviewing the draft audit report.
4.5 The lawyer directing the audit program is responsible for maintaining a controlled file of audit reports.
4.6 Recipients of audit reports are responsible for limiting their further distribution.

5. References/Appendices
5.1 ENV-800-G-PR-001 Environmental Compliance Auditing
5.2 ENV-800-G-PR-005: Environmental Compliance Audit Follow-Up/Action Plans

6. Procedure/Work Instruction
6.1 The overall goal of an audit report is to document the audit findings clearly and accurately and to communicate them to the appropriate levels of management in a timely manner.
6.2 After the audit, the audit team leader uses the findings presented at the audit's closing meeting, along with the

team's notes and collected evidence to draft the audit report at the direction of the audit team lawyer. The report usually contains;

- **An introduction**
- **Purpose, date, and scope of the audit (Any deviations from the planned scope should be noted.)**
- **Name, location, and brief description of the unit**

6.3 Approximately one month after receipt of the audit report, the unit manager submits the action plan to his/her immediate supervisor for approval, with a copy to the audit team leader.

6.4 The plan is discussed and modified, as needed, until it is approved by the unit manager's immediate supervisor.

6.5 The final action plan is given to the unit manager's immediate supervisor with copies to the division environmental *manager,* audit team leader, division environmental oversight coordinator, and Technical Services Center project manager.

6.6 The action plan is implemented by the unit. Throughout its implementation, the unit manager submits quarterly status reports to the unit manager's immediate supervisor, with copies to the division environmental manager, audit team leader, division environmental oversight coordinator, and Technical Services Center project manager.

6.7 The division environmental oversight coordinator checks implementation of the action plan during his/her periodic visits to the unit. Any exceptions or comments related to addressing audit findings are communicated to senior management.

6.8 Upon completion of the action plan, the unit manager writes a letter to his/her immediate supervisor, with copies to the division environmental manager, audit team leader, division environmental oversight coordinator, and Technical Services Center project manager to confirm that all issues have been addressed.

7. Records/Reports

7.1 Action Plan. The unit manager is responsible for preparation and implementation of the action plan.

7.2 Confirmation Letter. The unit manager is responsible for confirming in writing that the action plan has been fully implemented.

8. Revision History/Reason for Change

8.1 This is the first version of this document.

ENVIRONMENTAL COMPLIANCE AUDIT FOLLOW-UP/ACTION PLANS

Document Number: ENV-800-G-PR-005

1. Purpose

This document defines the procedure for environmental compliance audit follow-up and action plans.

2. Scope

This procedure applies to audits conducted as part of Asarco's internal environmental compliance audit program, managed by a Technical Services Center project manager at the direction of the Legal Department.

3. Definitions

3.1 None.

4. Responsibility

4.1 Unit staff is responsible for conducting the root cause analysis for the audit findings, as appropriate, as specified by AMS 410 b.

4.2 The unit manager is responsible for preparation and implementation of the action plan.

4.3 The unit manager's immediate supervisor is responsible for approving the final action plan prior to its implementation.

4.4 The, unit manager's immediate supervisor and the division environmental manager, are responsible for reviewing the unit's progress and following-up with the unit, as needed, until a letter is received confirming that all audit findings have been addressed.

4.5 The division environmental oversight coordinator is responsible for independently confirming that the approved action plan addresses each of the issues raised in the audit report and that the plan is being effectively implemented.

5. References/Appendices

5.1 ENV-800-G-PR-001: Environmental Compliance Auditing

5.2 ENV-800-G-PR-005: Environmental Compliance Audit Report Contents, Distribution, and Document Control

6. Procedure/Work Instruction

6.1 The overall goal of an action plan is to respond to all audit findings in a timely manner.

6.2 Upon receipt of the team leader's audit report, the unit manager develops an action plan proposal with the assistance of unit staff, the audit team members, the Legal Department, and/or the Technical Services Center, as needed. The action plan includes specific deliverables, responsibility assignments, root cause analysis (as appropriate), and a schedule.

- **Names of the audit team members and their locations**
- **Audit findings as described in ENV-800-G-PY-001**
- **General instructions for unit response and follow-up (see ENV-800-G-PR-005 Environmental Compliance Audit Follow-Up/Action Plans).**

6.3 As needed, the team leader has team members review, discuss, and revise the draft audit report for accuracy and clarity. Team members return any copies of the draft report and all work papers to the team leader. All audit-related documents should be marked "Confidential."

6.4 The team leader sends the draft audit report to the division environmental manager for review and comment.

6.5 Within one month of the audit's close, the team leader finalizes and distributes the audit report at the direction of the audit team lawyer. Usually, the audit report is issued to the unit manager, along with copies to the:

- **Unit manager's immediate supervisor**
- **Division environmental manager**
- **Division environmental oversight coordinator**
- **Lawyer responsible for direction of the audit program**
- **Technical Services Center project manager**

6.6 Recipients of the audit report limit its further distribution by making copies of the report only when absolutely necessary.

7. Records/Reports

7.1 Audit Report. The audit team leader issues the final report to a limited distribution. The lawyer responsible for direction of the audit program maintains a controlled file of audit reports.

8. Revision History/Reason for Change

8.1 This is the first version of this document.

Plea Agreement (U.S. v. BP Exploration (Alaska) Inc.)

ROBERT C. BUNDY
United States Attorney

DEBORAH M. SMITH
Deputy Chief
Environmental Crimes Section

TIMOTHY M. BURGESS
Assistant United States Attorney

SCOTT ELTRINGHAM
Trial Attorney
Environmental Crimes Section

JAMES OESTERLE
Special Assistant United States Attorney
Federal Building & U.S. Courthouse
Room 253
222 W. 7th Avenue, #9
Anchorage, AK 99513-7567
(907) 271-5071

IN THE UNITED STATES DISTRICT COURT
FOR THE DISTRICT OF ALASKA

UNITED STATES OF AMERICA Plaintiff, v. BP EXPLORATION (ALASKA) INC. Defendant.	)	No. A99-0141 CR (JKS) PLEA AGREEMENT

I. INTRODUCTION

A. This document contains the complete plea agreement (Agreement) between the United States of America (Government) and Defendant BP EXPLORATION (ALASKA) INC. (BPXA), a Delaware corporation doing business in Alaska. Other than this Agreement, no other agreement, understanding, promise or condition exists between the parties.

B. The parties expressly agree and acknowledge that this Agreement is entered into and controlled by Fed. R. Crim. P.1 l(e)(l)(C) and 1 l(e)(2). The Government and the Defendant understand and agree that: (a) if the Court accepts this Agreement, the Court is bound by the terms of the Agreement, and (b) if the Court does not accept this Agreement, either party may withdraw from the Agreement and the Defendant may withdraw its plea of guilty.

C. This Agreement does not limit the rights of any party to provide the Court or the United States Probation Office (Probation Office) with a full description of the Defendant's conduct, correct inaccuracies at any time, or speak at the time of sentencing consistent with the recommended provisions set forth in the Agreement. The parties agree to allocute at sentencing in favor of the joint recommendation regarding the appropriate sentence, as set forth in this Agreement.

D. This Agreement is limited to the United States Attorney's Office for the District of Alaska (United States Attorney's Office) and the Environmental Crimes Section of the Department of Justice, and does not bind other sections or divisions of the

Department of Justice or any other federal, state or local prosecutive or regulatory authority.

E. The Defendant agrees that it has been fully advised of its statutory and Constitutional rights, that it has been informed of the charges and allegations against it and the possible penalties, and that it understands them. The Defendant further agrees that it understands that by entering a plea of guilty as set forth below, it will be waiving certain statutory and Constitutional rights to which it is otherwise entitled.

F. The Government and the Defendant agree that Chapter 8 of the Federal Sentencing Guidelines Manual governs this case with regard to any payment of restitution, community service, and probation, consistent with the nature of this Rule 11 (e)(1)(C) Agreement; however, pursuant to U.S.S.G. §8C2. 1, Chapter 8 of the Federal Sentencing Guidelines Manual does not apply to the determination of an appropriate fine.

G. The Defendant understands and agrees that this Agreement does not confer any immunity on the Defendant or any of its officers, employees or directors for making false statements to the Government or giving untruthful testimony under an oath at any judicial proceeding, nor does this Agreement provide use immunity or derivative use immunity to the Defendant or any of its officers, employees or directors for any statements.

II. THE AGREEMENT

A. The Defendant agrees to:

1. Waive indictment and plead guilty to an Information charging it with one felony count of knowingly failing to immediately report the release into the environment of a hazardous substance, in violation of the Comprehensive Environmental Response, Compensation, and Liability Act (CERCLA), Title 42, United States Code, Section 9603(a).

2. Acknowledge responsibility for the acts and omissions constituting the crime alleged in the Information and constituting the factual basis for its plea of guilty, which facts are set forth below in Section VI of this Agreement.

3. Have its representative, an officer of BPXA, employed by and duly authorized by the Defendant's Board of Directors and with authority to speak for the Defendant, appear and enter the guilty plea and also appear for imposition of sentence.

4. Provide to the Government and the Court written evidence, in the form of a notarized resolution of its Board of Directors with both notary and corporate seals, certifying that the Defendant is authorized to waive its right to Indictment, to plead guilty to the Information in this case and to enter into and comply with all provisions of this Agreement. The resolution shall further certify that the President or Chief Executive Officer or his or her designee is authorized to take these actions and that all corporate formalities, including but not limited to approval by Defendant's directors, required for such authorization have been observed.

5. Forfeit its right to appeal any lawful sentence imposed in conformity with the provisions of this Rule 11(e)(1)(C) Agreement. The parties agree not to make any public statement regarding this Agreement or the Information, other than as required by law, prior to the filing of the Information.

6. Pay a fine of $500,000 as specified in paragraph IV.A of this Agreement.

7. Be placed on organizational probation for a probationary period of five years as specified in paragraph IV.C of this Agreement.

8. Provide full and complete cooperation to the United States in its continuing investigation of possible violations of federal law by any of the Defendant's employees, contractors, or agents as specified below. It is understood and agreed that the Government is not seeking and the Defendant is not agreeing to disclose legal or strategic advice or legal expertise rendered. The Defendant agrees to do the following:

a. Provide all reports of interviews prepared or conducted by its counsel (including in-house and outside counsel and criminal defense counsel) and its agents, and all notes of interviews containing factual information regarding possible violations of federal criminal environmental law relating to the disposal of hazardous waste at Endicott Island, Alaska, from 1992 to 1995. Portions of the reports of interviews which include counsel's mental impressions, opinions of counsel and advice of counsel may be redacted. The Defendant will advise the Government of any interviews that were not reduced to writing and have the interviewer generate a report of interview based on his/her recollection. In entering into this agreement, the Defendant understands that there may be a need to seek clarification of any report from its author by any appropriate means and to seek the testimony of any participant in any interview. The Government will only call an attorney for the Defendant as a witness in the event the Government determines in its sole discretion that securing adequate testimony through other witnesses, such as investigators, is not practicable or effective;

b. Provide all documents and portions thereof containing statements, including factual summaries, digests and/or chronologies prepared by its agents, employees or counsel regarding the subject matter defined in subparagraph 8(a);

c. Provide all relevant written factual information concerning the complete results of all internal and external investigations of these matters;

d. Provide full access to the Defendant's factual and technical experts, consultants and vendors and copies of all factual records and reports produced by same, or factual portions thereof, including those prepared for or at the direction of counsel. The Government will restrict any of its questioning, interviews or testimony of these individuals to factual information and documents;

e. Provide an updated and consolidated index for all documents produced pursuant to subparagraphs 8(a) through

8(d) of this Agreement and the subpoenas issued during the Government's investigation of BPXA;

f. Provide an updated and consolidated privilege log for all documents withheld from production pursuant to subparagraphs 8(a) through 8(d) of this Agreement and the subpoenas issued during the Government's investigation of BPXA under a claim of privilege, with an indication of which documents (or portions thereof) are not being provided and the justification. As part of the consolidated privilege log, or in a separate document, the Defendant will identify and enumerate all documents being withheld in part, or in their entirety, under a claim of the joint defense privilege or the common interest privilege and provide a general description of each document and its origin, without disclosure of the identity of the members of any joint defense privilege. The Defendant will further provide the United States with a general description including the number, length and type of documents withheld on the basis of the joint defense privilege;

g. The Defendant, and all counsels and consultants for the Defendant, shall withdraw from all joint defense agreements and all common interest agreements regarding the subject matter of this Agreement upon signing of this Agreement and shall immediately sever all such contractual agreements. The Defendant, and all counsels and consultants for the Defendant, agree not to enter any joint defense agreements or common interest agreements regarding the subject matter of this Agreement after the signing of this Agreement. The Defendant shall not produce information or notes of interviews without the consent of all parties to the interview if such interview is protected by a common interest agreement or joint defense privilege;

h. During the period of probation, and at all reasonable times and with reasonable prior notice by the Government as practicable, provide full access to its facilities, records, and employees including: (1) access to, and production and authentication of, all records regarding the subject matter defined in subparagraph 8(a); and (2) access to its officers and employees for the purposes of service of process, interviews, or arrest;

i. Make its employees available in the United States to participate in judicial proceedings, and the Defendant will make all reasonable efforts to ensure that its officers, employees, consultants, attorneys, vendors and experts cooperate with the investigation and disclose all information about their activities and those of others relating to the subject matter set forth in subparagraph 8(a), including advising them that: (1) they are encouraged to cooperate, including by agreeing to provide information to the Government, being interviewed by Government agents or attorneys, and testifying in official proceedings; and (2) the Defendant will grant liberal leave and pay reasonable transportation expenses to facilitate their cooperation with the Government.

9. In providing the factual information and factual documents identified in paragraph 8, and factual portions thereof, the Defendant and all its counsel affirmatively waive any attorney-client privilege or attorney work product doctrine that may apply to this information and documents. The Defendant and all its counsel agree that the United States may make any direct or indirect use, including release to third parties, of any information and documents provided under this Agreement, except that in the event the United States proposes to release confidential business information to any third party, the Defendant shall be provided prior notice and a reasonable opportunity to seek an appropriate protective order or other judicial relief. The Defendant shall provide to the United States an index of materials which it defines as confidential business information. The Defendant will not assert against the United States that any of the factual information or documents provided under this agreement remain subject to the attorney-client privilege or remain protected by the work product doctrine. The Government agrees not to argue that the fact that the Defendant has produced factual information and documents under this Agreement entitles the United States to attorney advice, legal strategy or legal opinions otherwise subject to the attorney-client privilege or the work product doctrine. The parties understand and agree that the United States may request additional documents from the Defendant as authorized by law. However, waivers required under this paragraph are not intended as general waivers of the attorney-client or work product privileges and apply only to the obligations defined in subparagraphs 8(a) through 8(i) of this Agreement, the subpoenaed documents, and information, interviews and documents voluntarily provided to date.

10. The Defendant agrees to develop, implement and maintain a Nationwide Environmental Management System (EMS) at BP Amoco Group facilities which are engaged in the exploration, drilling and/or production of oil, both onshore and offshore, in the United States of America and its territories, as described below. The EMS will be implemented and maintained at the following BP Amoco Group Business Units, which are owned, managed or operated by BPXA, BP Exploration & Oil, and/or Amoco Production Company. First, within Alaska, the participating Business Units will include Alaska Exploration, Central North Slope, Eastern North Slope, and Western North Slope. Second, within the contiguous United States, the participating Business Units will include Western U.S. Gas and MidContinent-Gulf Coast. Third, offshore the contiguous United States, the participating Business Units will include Gulf of Mexico-Deepwater Development, Gulf of Mexico-Deepwater Production, Gulf of Mexico-Exploration, and Gulf of Mexico-Shelf. The EMS will be consistent with the best environmental practices in order to effectively protect workers, the public and the environment, and to comply with all statutory and regulatory requirements. Under the EMS, policies and procedures will be implemented which are designed to ensure that the environmental performances of contractors and on-site service organizations, such as BPXA Shared Services Drilling, are adequately supervised. The Defendant agrees to assume all costs for the development, implementation and maintenance of the EMS during the period of probation, which will amount to approximately $15 million. Since BPXA's report of the illegal conduct to EPA in September 1995, approximately $5 million has been spent to implement the present environmental management system on Endicott Island and in Alaska.

a. BP America Inc., a Delaware corporation and Defendant's parent corporation, has agreed to (1) guarantee the availability of the funding necessary to develop, implement, maintain and monitor the EMS as described in this agreement and Attachment 1, (2) guarantee the performance of BPXA and BP Exploration & Oil in accord with this

Agreement and Attachment 1, and (3) cooperate fully in the development, implementation, maintenance and monitoring of the EMS, as described in this Agreement and Attachment 1.

b. BP Amoco Corporation, an Indiana corporation and a parent corporation of Amoco Production Company, has agreed to (1) guarantee the performance of Amoco Production Company in accord with this Agreement and Attachment 1, and (2) cooperate fully in the development, implementation, and maintenance of the EMS as described in this Agreement and Attachment 1.

c. The Defendant, and each of the corporations which manage, own or operate each of the individual Business Units, including BP Exploration & Oil and Amoco Production Co., shall agree and guarantee that the respective obligations to be assumed under paragraphs 10 through 12 and Attachment 1 attach to the assets or facilities subject to the EMS and shall pass to any successor-in-interest to each corporation, and shall be extended to any new Business Unit or new division engaged in the exploration, drilling and/or production of oil, provided that the assets or facilities of such Business Unit or new division are wholly owned, managed or operated by or affiliated with the Defendant, each of the aforesaid corporations or an affiliated corporation of them.

d. BP Amoco p.l.c., an English corporation and the controlling corporation of the BP Amoco Group of companies, including BPXA, BP Exploration & Oil and Amoco Production Co., (BP Amoco) may from time to time merge with or acquire companies or assets during the term of this Agreement. The Defendant acknowledges and agrees that it is one of the objectives of paragraphs 10 and 11 of this Agreement to establish a uniform EMS in all new assets or facilities acquired by BP Amoco wholly or partly engaged in the exploration, drilling and/or production of oil and within the United States of America and its territories during the term of this Agreement. Should BP Amoco merge with or acquire additional assets or facilities during the term of this Agreement which are wholly or partly engaged in the exploration, drilling and/or production of oil and within the United States of America and its territories, then the Defendant shall immediately identify those new assets or new facilities which are engaged in the exploration, drilling and/or production of oil within the United States of America and its territories and shall fund, develop, implement and maintain the EMS described in Attachment 1 of the Agreement for those assets or facilities, in accordance with the provisions and limitations of paragraphs 10 through 12 and Attachment 1, provided that the acquired assets are, after the acquisition, operated or wholly or partially owned corporations controlled by BP Amoco or its successor-in-interest. Board resolutions of the corporate parents of any corporations owning such assets or facilities, and guarantee letters, similar in form and content to those delivered by the corporations presently owning such assets or facilities of the Business Units listed in Attachment I of the Plea Agreement, shall be delivered by the Defendant to the United States Attorney's Office for the District of Alaska and the United States Probation Office within thirty (30) days of the formal completion of such acquisition.

e. The Defendant agrees to provide the United States Attorney's Office for the District of Alaska and the United States Probation Office with immediate notice of any name change, business reorganization, sale or purchase of assets, divestiture of assets, or similar action affecting the Business Units subject to this Agreement or Defendant's performance of this Agreement. The Defendant shall not through a change of name, business reorganization, sale or purchase of assets, divestiture of assets, or similar action, seek to avoid the obligations and conditions set forth in this Agreement.

f. This Agreement, together with all of the obligations and terms hereof, shall inure to the benefit of and shall bind assignees, successors-in-interest, or transferees of the Parties.

g. No change in ownership or corporate or other legal status including, but not limited to, any transfer of assets or property, shall alter BPXA's responsibilities under this Plea Agreement, except, however, in the event BPXA no longer has any ownership interest or operational control at a facility other than Endicott Island and Satellite Drilling Island, it may notify the Environmental Protection Agency in writing, providing supporting documentation of the transaction, and, thereafter, will no longer be obligated to perform the requirements of Section II.A, paragraphs 10 through 12, and Section IV.D, paragraph 3, of this Plea Agreement and Attachment 1 thereto at that facility. No later than thirty (30) days prior to sale or transfer of ownership or operation of any Endicott Island or Satellite Drilling Island facility, BPXA shall give written notice of this Plea Agreement to each purchaser or successor-in-interest. Upon such sale or transfer, BPXA shall provide a copy of this Plea Agreement to each such purchaser or successor-in-interest. The sale or transfer of ownership or operation of any Endicott Island or Satellite Drilling Island facility will not relieve BPXA of its obligations regarding the Endicott Island and Satellite Drilling Island facility under this Plea Agreement. BPXA shall notify in writing the United States Department of Justice and the Environmental Protection Agency of such purchaser or successor-in-interest at least thirty (30) days prior to any sale or transfer of any Endicott Island and/or Satellite Drilling Island facility.

11. The Defendant agrees to ensure that the incentive bonus and shared savings programs offered to the employees, contractors, contractor employees and agents engaged in oil exploration, drilling and production at BP Amoco Group facilities in the United States and its territories do not encourage behavior which could result in operational risks or environmental harm. The improved incentive bonus programs and shared savings programs will be implemented and maintained at BP Amoco Group Business Units owned, managed or operated by BPXA, BP Exploration & Oil, and/or Amoco Production Company, as listed in paragraph 10. Waste minimization strategies adopted to avoid costs will not be rewarded unless those strategies are wholly consistent with all applicable statutes and regulations. All such incentive bonus and shared savings programs will be regularly evaluated to ensure that their existence does not cause additional environmental risks. Should any new assets or facilities be acquired by BP Amoco as described in subparagraphs 10(c) and (d), their incentive bonus and shared savings

programs shall be subject to the same provisions as in this section.

12. The Defendant agrees to the appointment of a court-appointed monitor during the probationary period to monitor development, implementation and maintenance of the EMS and the design and supervision of the incentive bonus and shared savings programs. The Defendant agrees to assume all costs and expenses associated with the employment of the monitor.

B. In exchange for the Defendant's full and complete compliance with the terms of this Agreement, the Government acknowledges and agrees that:

1. Following the Defendant's report of the criminal conduct to the Environmental Protection Agency, BPXA did the following: improved its waste management procedures at Endicott Island; enhanced its environmental training for BPXA employees and contractors; discontinued the practice of utilizing used solvents and used oil to protect the outer annuli of wells from freezing; commissioned a waste management audit on Endicott Island; conducted a North Slope-wide multi-media environmental compliance audit; qualified for certification pursuant to the International Organization for Standardization (ISO) Environmental Management Systems-Specification with Guidance for Use (also known as ISO 14001 certification); reorganized its Health, Safety and Environmental (HSE) sections within the business organization to increase corporate accountability for HSE functions and to ensure compliance with environmental statutes and regulations; and modified its incentive bonus program to ensure that employees did not have an incentive to cut costs at the expense of environmental compliance and safety.

2. During the Government's investigation, BPXA took the following steps to cooperate: provided a summary of its initial internal investigation, including notes of interviews; provided extensive documentation; provided unrestricted access to the facilities and drilling operations on Endicott Island, at Prudhoe Bay and to the BPXA Anchorage offices; provided unrestricted access to paper files and computer files on Endicott Island and in BPXA Anchorage offices; provided a document-by-document inventory of documents produced; provided access to expert witnesses; made employees stationed in foreign countries available to Government investigators, and paid for the employees' transportation to Anchorage. After a period of negotiation, BPXA provided additional notes of employee interviews taken by outside counsel during the later stages of the internal investigation, and provided a limited waiver of the attorney-client and work product privileges.

3. In recognition of the corrective measures adopted by the Defendant subsequent to the offense, the development, implementation and maintenance of the Nationwide Environmental Management System, and certain steps taken to cooperate with the government's investigation, the Government agrees to forego charging in an information or indictment criminal violations of the Resource Conservation and Recovery Act, Title 42, United States Code, Section 6928(d)(2).

4. The Government will not seek additional criminal prosecution in the District of Alaska against the Defendant, or any other affiliated or related corporate entity, for environmental criminal violations and general criminal violations (1) relating to waste disposal and rig operations at Rig 15 from 1992 to 1995, or (2) based on any matters that were the subject of the Government's criminal investigation involving the Defendant and were known to the Government at the time of this Agreement. This Agreement does not limit the Government's right to prosecute any offenses based on facts of which it was unaware as of the date of this Agreement. The parties understand that this Agreement does not apply to any individuals, including but not limited to present and former employees, officers, agents and contractors. The parties further understand that this Agreement applies only to federal criminal charges and only binds the United States Attorney's Office for the District of Alaska and the Department of Justice Environmental Crimes Section. The Defendant has discussed this Agreement with its attorneys and understands that nothing contained in this Agreement is meant to limit the rights and authority of the United States to take further civil or administrative action against the Defendant or any affiliated or related corporation, including but not limited to, any listing and debarment proceedings to restrict rights and opportunities of the Defendant to contract with or receive assistance, loans and benefits from United States government agencies.

III. PENALTIES

The maximum penalty for the count in the Information is a fine of $500,000 or twice the gross gain or loss resulting from the unlawful conduct, Title 18, United States Code, Section 3571; from one to five years probation; and a special assessment of $400, Title 18, United States Code, Section 301 3(a)(2)(B).

IV. RECOMMENDATIONS AS TO SENTENCE

Pursuant to Fed. R. Crim. P.1 l(e)(l) and (2), the parties agree that the following sentence is appropriate and should be imposed in this case:

A. The Defendant shall pay the maximum fine of $500,000. The fine will be paid on or before the date that the Court imposes sentence.

B. The Defendant shall pay a special assessment of $400 to the Victim's Assistance Fund, pursuant to Title 18, United States Code, Section 3013(a)(2)(B), on or before the date that the Court imposes sentence.

C. The Defendant shall be placed on organizational probation for a period of five years.

D. In addition to the Court's standard conditions of probation and in addition to conditions that may be set by the Probation Office, the terms of probation shall include the following specific conditions:

1. The Defendant shall commit no further violations of environmental laws, regulations or permits of the United States, including those for which primary enforcement has been delegated to a state.

2. Any environmental violation voluntarily and timely reported to appropriate representatives of the United States Government by BPXA, as provided in its EMS and in accord with all the provisions of the 1991 Department of Justice voluntary disclosure policy (also known as "Factors in Decisions on Criminal Prosecutions for Environmental Violations in the Context of Significant Voluntary Compliance or Disclosure Efforts by the Violator"), shall not be considered by prosecutors to be a probation violation or to be subject to sanctions under this Agreement. The determination of whether the disclosure is timely and voluntarily reported shall be within the sole discretion of the Government. The Government's position on whether a subsequent violation is an appropriate basis for a probation violation does not bind the United States Probation Office or the Court. Nothing herein shall prohibit the United States from proceeding administratively, civilly, or criminally against BPXA for any alleged environmental violation.

3. The Defendant shall develop, implement and maintain a permanent Nationwide Environmental Management System, as described in paragraph 10 and Attachment 1, "BP Amoco Group Environmental Management System," which is incorporated by reference, at all BP Amoco Group Business Unit facilities engaged in the exploration, drilling and/or production of oil, both onshore and offshore, in the United States of America and its territories. Attachment I details the EMS objectives and required supervision, structure, training, audits, and reports to the Government and Court. Paragraph 10 of this Agreement contains requirements relating to the acquisition of Business Units and such terms shall apply throughout this section.

a. The EMS at each such Business Unit facility will establish that: all environmental and related operational risks have been identified; such risks are being appropriately managed and potential risks are avoided; all federal, state and local environmental laws, regulations, and permit requirements are being adhered to; appropriate policies, programs and procedures are in place; organizational responsibilities are clearly defined, understood and implemented; Business Units', contractors' and on-site service organizations' environmental quality control assurance and verification systems are in place, as determined by appropriate self-policing and third-party audits; company operations, including contractor operations and on-site service provider operations, do not present actual risks to the environment. The Defendant shall ensure that the environmental compliance program is diligently enforced by the officers and managers of the Defendant and the Business Units.

b. The Defendant shall be responsible for all costs associated with the development, implementation, maintenance and monitoring of the EMS.

c. As part of the EMS, the Defendant shall ensure that all BP Amoco Group Business Unit incentive bonus and shared savings programs, as described in paragraph 10 and Attachment 1, offered to employees, contractors, contractor employees, and agents engaged in the exploration, drilling and production of oil are designed to discourage behavior which could result in operational risks or environmental harm. All such incentive bonus and shared savings programs shall be regularly evaluated by the Defendant to ensure that their existence and terms do not cause environmental risks.

d. During the period of probation, a court-appointed monitor shall monitor the development, implementation and maintenance of the EMS and the design and supervision of the incentive bonus and shared savings programs, in accord with Attachment 1. The court-appointed monitor will be approved by the Court, United States Attorney's Office, Probation Office, and Environmental Protection Agency. The Defendant will assume all costs and expenses associated with the employment and expenses of the monitor. The Defendant and the Business Units agree to provide the monitor, immediately upon request, with unrestricted access to all facilities, employees, documents and computers, relevant to monitoring of the EMS and to facilitate access to contractors and contractor employees. The Court, Probation Office and United States Attorney's Office will establish the reporting requirements. All reports and draft reports issued by the monitor will be delivered to the Court, Probation Office and United States Attorney's Office prior to or simultaneously with delivery to the Defendant.

V. ELEMENTS

The elements of the offense described in the Information, to which the Defendant agrees to plead guilty, are as follows: (1) a hazardous substance; (2) in an amount equal to or greater than the reportable quantity; (3) was released into the environment; (4) from a facility; and (5) the Defendant, who was a person in charge of the facility; (6) failed to immediately notify the appropriate agency of the United States as soon as the Defendant became aware of the release. Title 42, United States Code, Section 9603(a).

[MATERIAL OMITTED]

BPXA acknowledges that from January 1993 through September 1995, it exercised inadequate oversight and supervision of the Health, Safety and Environmental (HSE) functions of BPXA's Shared Services Drilling and Doyon Drilling; that inadequate audits and audit protocols of BPXA HSE functions were in place; that inadequate resources were dedicated to ensuring appropriate environmental planning, oversight and self-policing; and there was inadequate coordination within the BPXA business organizations on HSE matters.

Dated this ___ day of September, 1999.

FOR THE UNITED STATES:

ROBERT C. BUNDY
UNITED STATES ATTORNEY

DEBORAH M. SMITH
DEPUTY CHIEF
ENVIRONMENTAL CRIMES SECTION
U.S. DEPARTMENT OF JUSTICE

TIMOTHY M. BURGESS
ASSISTANT U.S. ATTORNEY

SCOTT ELTRINGHAM
TRIAL ATTORNEY
ENVIRONMENTAL CRIMES SECTION

JAMES OESTERLE
SPECIAL ASSISTANT U.S. ATTORNEY

FOR BPXA:

CHRIS J. PHILLIPS
VICE PRESIDENT
BP EXPLORATION (ALASKA) INC.

CAROL E. DINKINS
VINSON & ELKINS
COUNSEL FOR BP EXPLORATION (ALASKA) INC.

DAVID B. BUKEY
BUKEY & BENTLEY
COUNSEL FOR BP EXPLORATION (ALASKA) INC.

JEFFREY M. FELDMAN
FELDMAN & ORLANSKY
COUNSEL FOR BP EXPLORATION (ALASKA) INC.

[MATERIAL OMITTED]

ATTACHMENT 1
PLEA AGREEMENT

United States V. BP Exploration (Alaska) Inc.
District of Alaska

BP AMOCO GROUP ENVIRONMENTAL MANAGEMENT SYSTEM

A. Objectives

BP Amoco Group will develop an Environmental Management System (EMS) for each Business Unit involved in the exploration, drilling and/or production of oil in the United States of America and its territories. Each EMS will be consistent with the best environmental practices in order to effectively protect workers, the public and the environment, and to comply with all statutory and regulatory requirements. Each EMS will also specify how the environmental performance of contractors and on-site service providers such as Shared Services Drilling in BP Exploration (Alaska) Inc. (BPXA), will be adequately supervised by BP Amoco Group in accord with the objectives and key elements described below. The term environment, as used in this document, includes the surroundings in which an organization operates, including land, water, air, flora, fauna, natural resources, humans, and their interrelation.

BP Amoco Group, as used in this document, means all BP Amoco p.l.c. companies, subsidiaries or affiliated business entities engaged wholly or partially in the exploration, drilling and/or production of oil in the United States of America and its territories. The EMS will be developed, implemented and maintained at all Business Units engaged wholly or partially in the exploration, drilling and/or production of oil in the United States and its territories, which BP Amoco Group manages, operates or in which it holds a controlling ownership interest. As of September 23, 1999, the BP Amoco Group consisted of BPXA, BP Exploration & Oil, and Amoco Production Company, which manage, operate or own a controlling interest in the Business Units listed below.

The EMS will be implemented and maintained at the following BP Amoco Group Business Units. First, within Alaska, the participating Business Units will include Alaska Exploration, Central North Slope, Eastern North Slope and Western North Slope. Second, within the contiguous United States, the participating Business Units will include Western U.S. Gas and

MidContinent-Gulf Coast. Third, offshore the contiguous United States, the participating Business Units will include Gulf of Mexico-Deepwater Development, Gulf of Mexico-Deepwater Production, Gulf of Mexico-Exploration and Gulf of Mexico-Shelf.

BP Amoco Group will establish three Compliance Committees to evaluate the existing EMS at each relevant Business Unit (BU), prepare an improved EMS for each of these BUs in accord with the key elements and objectives described below, report to the Environmental Protection Agency, and implement the EMS at each BU in the manner described below. The Compliance Committees will also manage and oversee environmental requirements that fall outside the jurisdiction of any one BU and will ensure that cross-cutting environmental requirements are appropriately coordinated among BUs. The Compliance Committees will also ensure that procedures outlined within the environmental management systems are standardized whenever practical, so it will not be necessary for BP Amoco Group employees transferred from one BU to another to learn about many different procedures at their new locations. The three Compliance Committees will oversee all Alaska BUs, and the onshore and the offshore BUs of the contiguous United States which are engaged in the exploration, drilling and production of oil.

The key elements of each BU EMS will be derived from the following sources:

· BP Amoco p.l.c. and BP Amoco Group principles and requirements;
· The International Organization for Standardization (ISO) specification for environmental systems, "Environmental Management Systems-Specification with Guidance for Use," 1996, also known as ISO 14001 standard;
· U.S. Department of Justice Voluntary Disclosure Policy, "Factors in Decisions on Criminal Prosecutions for Environmental Violations in the Context of Significant Voluntary Disclosure Efforts by the Violators," 1991,
· U.S. Environmental Protection Agency's (EPA) "Incentives for Self-Policing: Discovery, Disclosure, Correction and Prevention of Violations," 1995;
· EPA's "Implementation Guide for the Code of Environmental Principles," 1997;
· EPA's National Enforcement Investigations Center "Compliance-Focused Environmental Management System-Enforcement Agreement Guidance," 1997; and
· There will be enhanced emphasis on compliance assurance, pollution prevention and adequate contractor and on-site service provider supervision.

Each BU EMS will establish that:

· All environmental and related operational risks have been identified;
· Risks are being appropriately managed and potential risks are avoided;
· All federal, state and local environmental laws, regulations, and permit requirements are being adhered to;
· Appropriate policies, programs, and procedures are in place;
· Organizational responsibilities are clearly defined, understood, and implemented;
· Appropriate environmental control quality assurance, and verification systems are in place through appropriate self-policing and third-party audits of BP Amoco Group, contractor and on-site service provider operations;
· Company operations, including contractor operations and on-site service provider operations, do not present unreasonable actual risks to the environment.

Each BU EMS will be consistent with the following ISO 14001 principles:

· Line Management and Responsibility: BP Amoco Group BU line management will be directly responsible for protecting workers, the public and the environment.
· Clear Roles and Responsibilities: Clear and unambiguous lines of authority and responsibility for ensuring environmental protection will be established and maintained at all organizational levels, and additionally for contractors and on-site service providers.
· Competence Commensurate with Responsibilities: BP Amoco Group personnel will possess the experience, knowledge, skills, and abilities that are necessary to discharge their responsibilities.
· Balanced Priorities: BP Amoco Group will effectively allocate resources for environmental programs and operations. Protecting workers, the public and the environment will be a priority at the BP Amoco Group BUs whenever activities are planned and implemented.
· Identification of Environmental Hazards: Before any task is begun, BP Amoco Group BU management will evaluate the associated environmental hazards.
· Operational and Hazard Control: Requirements and standards for environmental performance will be established which, if properly implemented, will adequately ensure that workers, the public and the environment are protected from adverse consequences. BP Amoco Group will include administrative and engineering controls to prevent and mitigate environmental hazards and to achieve other environmental goals, for example, pollution prevention.
· Checking and Corrective Action: BP Amoco Group will routinely monitor its environmental performance, including its compliance with legal requirements and conformance to environmental management system requirements. BP Amoco Group will take actions as necessary to correct any problems identified and to prevent their recurrence.
· Continual Improvement: Top management will periodically evaluate the suitability, adequacy and effectiveness of the environmental management system and take actions as necessary to ensure its ongoing suitability, adequacy and effectiveness.

B. Environmental Management System Manual - Key Elements

Each BU EMS will cover, at a minimum, the following 12 key elements that will be described in an Environmental Management System Manual.

Key Element 1
Management Policies and Procedures

· BP Amoco Group's Environmental Policy. The policy will clearly communicate BP Amoco Group's commitment to environmental performance, including compliance with all applicable federal, state and local environmental statutes, regulations, enforceable agreements, and permits (environmental requirements). The policy will also clearly communicate BP Amoco Group's commitment to pollution prevention and continual improvement of the environmental management systems.
· Site-Specific Environmental Policies and Standards

1) Compilation of general principles, policies, rules, and procedures for site-specific environmental practices.
2) Includes process for developing, approving, and communicating standard operating practices for activities having potentially adverse environmental or regulatory compliance impacts.
3) Clearly identifies organizational responsibilities for maintaining compliance with environmental requirements, including required reporting to regulatory agencies.
4) Includes ongoing means of communicating environmental issues and information to all organization personnel, on-site service providers, and contractors, and for receiving and addressing their concerns.
5) Establishes and describes processes to ensure sustained interaction with regulatory agencies, and within the organization (e.g., between the various divisions, contractors, on-site service providers and the environmental department) regarding compliance with environmental requirements.

Key Element 2
Organization, Personnel, and Oversight of EMS

· Describes organizationally, how the EMS is implemented and maintained.
· Includes organization charts that identify units, line management and other individuals having responsibility for compliance with environmental requirements.
· Identifies and defines duties, roles, responsibilities and authorities of key environmental program personnel in implementing and sustaining the EMS.
· Ensures that specified responsibilities and accountabilities are consistently reflected in system documentation.
· Describes process to conduct and document periodic needs assessments and ensure adequate resources are provided for environmental compliance, including staffing, funding, equipment and external support.

Key Element 3
Accountability and Responsibility

· Specifies accountability and responsibilities of BU management for environmental protection practices, compliance, required reporting to regulatory agencies and corrective actions implemented in their area(s) of responsibility.
· Specifies accountabilities and responsibilities of BU management to adequately supervise the compliance with environmental, safety and health requirements by BU contractors and on-site service providers.
· Describes potential consequences for departure from specified operating procedures, including liability for criminal/civil/administrative penalties imposed as a result of noncompliance with environmental requirements.
· Measures employee and manager performance in area of environmental performance. Environmental performance will be an important aspect of performance appraisal and promotion of those staff for whom such activities form part of their normal activities. Leading (input) measures of environmental performance will be used, such as the number of compliance audits performed, percentage of audit findings closed within 30, 60, 90 days, percentage of compliance awareness training completed, number of waste minimization opportunities identified, number of spill reduction opportunities identified, and number of employees participating on HSE committees.
· BU employee, contractor and on-site service provider incentive bonus programs and shared savings programs will reward positive environmental performance. There will be an emphasis on utilizing leading (input) metrics to measure performance. A loss or diminution of bonus pay, or shared savings or assessment of any financial penalty will not be based upon lagging (output) metrics, which may discourage the accurate reporting of non-compliance with environmental requirements. For example, the number of environmental violations, regulatory incursions or spills will not be used as a metric or target.

The targets and performance measurements will emphasize continual improvement and accurate reporting. For example, in addition to leading metrics described above, a year-on-year improvement in overall air or water emissions per unit of production could be rewarded.

· Incentive bonus and shared savings programs for BU employees and for contractors or on-site providers will be designed to avoid encouraging behavior which could be anticipated to result in operational risks or environmental harm. Waste minimization strategies adopted to avoid costs will not be rewarded, unless those strategies are wholly consistent with all applicable statutes and regulations. All such programs will be regularly evaluated to ensure that their existence does not cause additional environmental risks; should modifications be required in order to preclude any identified risks then these will be made as soon as possible. The impact of employee decisions or actions taken to reduce the risk of high-potential environmental incidents will not be considered in the calculation of lifting costs and production levels for inclusion in bonus and shared savings programs. For example, the cost of remedial cementation during the drilling of a well (or of well repairs or well abandonment for environmental reasons) will not be considered in such calculations.

· Describes process to measure and supervise contractor and on-site service provider environmental performance. On-site service provider and contractor self-inspections, training sessions, and other leading metrics described above will be used, in addition to information tracking spills and injuries. Results of this process will be used to take corrective action as needed.

· Implements a process to ensure that contractor interface documents are implemented as written.

· Establishes a process to ensure that contractors and on-site providers are accurately reporting spills, environmental incidents and other environmental performance data.

· Ensures the objectivity of contractor and on-site service providers personnel performing audits or other environmental performance self-assessments.

Key Element 4 Environmental Requirements

· Describes process for identifying, interpreting and effectively communicating environmental requirements to affected BU personnel, on-site service providers and contractors, and ensuring that BU activities conform to those requirements.

· Specifies procedures for prospectively identifying and obtaining information about changes and proposed changes in environmental requirements, and incorporating those changes into EMS.

Key Element 5 Assessment, Prevention, and Control

· Identifies an ongoing process for assessing operations, for the purposes of preventing and controlling releases, ensuring environmental protection, and maintaining compliance with statutory and regulatory requirements. This section will describe monitoring and measurements, as appropriate, to ensure sustained compliance. It shall also include identifying operations and waste streams and discharges or emissions which may be causing, or may lead to: (1) releases of hazardous waste or other pollutants to the environment, or (2) a threat to human health or the environment. Finally, the process shall include performing root-cause analysis of identified problems in order to develop remedies and prevent recurring issues.

· Describes process for identifying operations and activities which could result in adverse environmental impacts, threats to human health or safety, and/or noncompliance with environmental requirements, where documented standard operating practices need to be developed.

· Describes process for ensuring input of environmental requirements (and concerns) in planning, design, and operation of ongoing, new, and/or changing buildings, processes, equipment, maintenance activities and product.

· Describes a system for conducting and documenting routine, objective, self-inspections by department supervisors and trained staff of BU, contractors and on-site service providers, especially at locations identified by the process described above.

· Describes a system for conducting and documenting systematic, objective, periodic review of facility operations and practices related to meeting environmental requirements by qualified, independent auditors. Develops and implements a systematic audit approach that employs consistent methods, site selections, protocols and procedure, updated as the law and regulations require.

· Describes and demonstrates top management commitment to compliance auditing program.

· Ensures adequate resources provided for compliance auditing program and assigns responsibilities for the compliance auditing program; ensures pre-planning of systematic audits and coordination of audits.

· Describes process for ensuring the qualifications and objectivity of auditors.

· Describes process that ensures the results of audits are reported to BU top management.

· Describes process that ensures the results of audits are reported to relevant contractors and on-site providers.

· Describes compliance auditing program which includes quality assurance component whereby an independent objective, qualified third-party organization audits the BU compliance audit program, reporting directly to BU top management at least once every two years. The third-party audit will determine, among other issues, whether there is adequate coordination among BUs and Compliance Committees and appropriate standardization of EMS procedures.

Key Element 6
Environmental Incident and Noncompliance Investigations

· Describes standard procedures and requirements for reporting of incidents that threaten human health or the environment and/or noncompliance with environmental requirements by BU, contractors and on-site service providers.
· Establishes procedures for investigation of potential violations or noncompliance with environmental requirements by BU, contractors and on-site service providers.
· Describes a system for development, tracking, and effectiveness verification of corrective and preventative actions taken by BU, contractors and on-site service providers.
· Implements process for contractor and on-site service provider employees to notify BU management without fear of reprisal regarding environmental compliance issues and concerns, for example, a phone hot-line.
· Each of these procedures shall specify self-testing of such procedures, where practicable.

Key Element 7
Environmental Training, Awareness, and Competence

· Identifies specific education and training required for BU personnel, contractors and on-site service providers, as well as process for documenting, evaluating and improving training provided.
· Describes program to ensure that BU employees, contractor employees and on-site service provider employees are aware of BU environmental policies and procedures, environmental requirements, and each employee's responsibilities within the environmental management system.
· Describes program for ensuring that personnel responsible for meeting and maintaining compliance with environmental requirements are competent on the basis of appropriate education, training and/or experience.
· Describes process for tracking BU employee, contractor and on-site provider questions about interpretation of environmental regulations. Such system feedback will provide valuable insight regarding the effectiveness of training and clarity of environmental procedures.

Key Element 8
Planning for Environmental Matters

· Describes how environmental planning will be integrated into other plans developed by the organization, as appropriate (e.g., capital improvements, training, maintenance).
· Requires establishing written goals, objectives, and action plans by operating organizational units with environmental responsibilities, as appropriate, including those for contractor operations conducted at the facility and on-site service providers, and how specified actions will be tracked and progress reported.

Key Element 9
Maintenance of Records and Documentation

· Identifies the types of records developed in support of the EMS (including self-audits, third-party audits and reports, audit working papers, and correspondence, including emails, about audits), who maintains them and where, and protocols for responding to inquiries and requests for release of information.
· Specifies the data management systems for any internal waste tracking, environmental data, and hazardous waste determinations.

Key Element 10
Pollution Prevention Program

· Describes an internal program for preventing, reducing, recycling, reusing and minimizing waste and emissions, including procedures to encourage material substitutions consistent with environmental requirements. Also includes mechanisms for identifying candidate materials to be addressed by program and tracking progress.

Key Element 11
Continuing Program Evaluation and Improvement

· Describes program for periodic (at least annual) evaluation of the EMS, including incorporating the results of the assessment into program improvements, revisions to the manual, improved training and communicating findings and action plans to affected employees, on-site service providers and contractors.

· Describes program for periodic (at least every two years) third-party evaluation of the EMS including incorporation of the results of the assessment into program improvements, revisions to the manual and communicating findings and action plans to affected employees, on-site service providers and contractors.

Key Element 12
Public Involvement/Community Outreach

· Describes a program for ongoing community education and involvement in the environmental aspects of the organization's products, activities and services and to promote general environmental awareness.

C. Establishment of Corporate Compliance Committee and Chair

BP Amoco Group will appoint three Corporate Compliance Committees (Committees) to oversee the environmental performance of all relevant BUs in the United States. Each Committee will have approximately 10 members, including environmental, engineering, drilling and operating personnel, BP Amoco Group corporate legal and environmental staff, and an independent, third-party consultant with expertise in designing environmental management systems. One committee will oversee all BUs involved in the exploration, drilling and production of oil onshore and offshore in Alaska; the second committee will oversee all BUs involved in the exploration, drilling and production of oil onshore in the contiguous United States and the third committee will oversee all BUs involved in the exploration, drilling and production located offshore the contiguous United States. The members of the Alaska Corporate Compliance Committee, including the independent third-party environmental consultant, must be submitted for approval to the U.S. Attorney's Office for District of Alaska and EPA Region 10 within two months of the execution of the plea agreement.

Each Committee will appoint a chair of the Committee. The chair's responsibility will be to ensure that the Committee meets all responsibilities and duties described below and to submit the reports and manuals described below to EPA Region 10 in a timely manner.

All costs related to the operation of Compliance Committees and the review, development, communication, audit and analysis of each BU Environmental Management System and Manual will be borne solely by BPXA, and guaranteed by BP America Inc.

D. Compliance Committee Responsibilities

1) Existing Environmental Management System Evaluation

Each-Committee will evaluate the current status of the environmental management system within each BU in order to:

· Compare the current BU environmental management systems to the EMS objectives and 12 key elements described above and determine what, if anything, is needed to meet these objectives and key elements;
· Identify strengths and weaknesses of the BU's current environmental management system;
· Evaluate the effectiveness of the current management system to effectively address the environmental issues and risks that the BU will face, now and in the future.

2) Preparation of Improved BU Environmental Management System

Each Committee will prepare and finalize an Environmental Management System for each BU that includes the best practicable methods for tracking, documenting and implementing environmental requirements, and clear departmental accountability for environmental and operational compliance. The Committees will prepare an EMS Manual for each BU that is organized to clearly address each of the key elements described above. The Committees will determine the most effective format for delivery of the manuals, including a computerized format, and insure that the current version of each BU's manual is always available to all of that BU's employees, contractors and on-site service providers.

3) Report to the Environmental Protection Agency About Endicott EMS

The Alaska Committee will complete its evaluation of the existing environmental management system on Endicott within 60 days following appointment and approval of the Committee membership. The first draft of the Environmental Management System Manual will be completed and submitted to EPA Region 10 for review within six months following the evaluation. EPA will review and comment upon the manual within 60 days. The Committee will finalize the manual after receipt of EPA comments and submit them within two months to EPA Region 10 for formal approval.

All elements of the Endicott EMS will be implemented within 24 months of the date the Endicott EMS Manual is formally approved. Within 18 months of the date that the Endicott EMS is fully implemented, the Alaska Committee will arrange for, fund and complete an independent, third-party EMS compliance audit, the purpose and scope of which shall be to evaluate the Endicott EMS to ensure its continuing suitability, adequacy and effectiveness and to recommend any changes appropriate to achieve the objectives of the program. The third-party EMS system auditors will be certified by the American National Standards Institute-Registrar Accreditation Board or will have comparable credentials and experience in performing EMS audits. The independent auditors and the scope of the audit will be pre-approved by EPA Region 10 prior to beginning the audit. The initial report of the auditors will be submitted simultaneously to EPA Region 10 and BP Amoco Group, along with audit working papers and correspondence related to the audit, including e-mails.

4) Report to the Environmental Protection Agency on EMSs Other than Endicott

The Committees will report to EPA Region 10 on the status of each BU EMS and whether it meets the objectives and key elements described above. A copy of each BU EMS Manual will also be submitted. The report and manual will be submitted to EPA Region 10 within two years of the execution of the plea agreement. EPA Region 10 will then forward the report and manual to the appropriate EPA Region.

5) Coordinate and Standardize Environment Management Systems

The Committees will jointly manage and oversee environmental requirements that fall outside the jurisdiction of a specific BU or Committee and will ensure that cross-cutting environmental requirements are appropriately coordinated among BUs. The Committees will also ensure that procedures outlined within each BU environmental management system are standardized whenever practical, so it will not be necessary for BP Amoco Group employees transferred from one BU to another to learn many different procedures at each location. Each Committee will be required to arrange for an independent, objective, qualified third-party organization audit, to determine if adequate coordination and standardization has occurred. This audit result will be reported directly to top management of BP Amoco Group and will be conducted at least every two years.

SELECT BIBLIOGRAPHY

General

Benforado, Jay et al. *Another Parallel Track?*, ENVTL. F., Nov./Dec. 1999, at 46.

Boergers, Kathleen. *The EPA's Supplemental Environmental Projects Policy*, 26 ECOLOGY L.Q. 777 (1999).

Cairncross, Frances. GREEN INC. (Island Press 1995).

____. Costing the Earth (Harvard Bus. School Press 1991).

Carey, Christopher D. *Negotiating Environmental Penalties: Guidance on the Use of Supplemental Environmental Projects*, 44 A.F. L. REV. 1, 18 (1998).

Carr, Donald A., ed. ENVIRONMENTAL CRIMINAL LIABILITY: AVOIDING AND DEFENDING ENFORCEMENT ACTIONS (BNA 1995).

Cooney, John F. et al. *Criminal Enforcement of Environmental Laws: Parts I-III*, 25 ELR 10459, 10525 & 10600 (Sept., Oct., and Nov. 1995, respectively).

Dana, David A. *The Uncertain Merits of Environmental Enforcement Reform: The Case of Supplemental Environmental Projects*, 1998 WIS. L. REV. 1181.

Droughton, Laurie. *Supplemental Environmental Projects: A Bargain for the Environment*, 12 PACE ENVTL. L. REV. 1141 (1995).

Friedman, Frank B. PRACTICAL GUIDE TO ENVIRONMENTAL MANAGEMENT (8th ed., Environmental Law Inst. 2000).

Harrell, Martin. *Organizational Environmental Crime and the Sentencing Reform Act of 1984: Combining Fines With Restitution, Remedial Orders, Community Service, and Probation to Benefit the Environment While Punishing the Guilty*, 6 VILL. ENVTL. L.J. 243 (1995).

Herman, Steven A. *EPA's Revised Supplemental Environmental Projects Policy Will Produce More Environmentally Beneficial Enforcement Settlements*, NAT'L ENVTL. ENFORCEMENT J., July 1995, at 9.

_____. Memorandum to Regional Administrators Regarding Issuance of Final Supplemental Environmental Projects Policy (Apr. 10, 1998).

Hoffman, Andrew J. COMPETITIVE ENVIRONMENTAL STRATEGY: A GUIDE TO THE CHANGING BUSINESS LANDSCAPE (Island Press 2000).

Lawrence, Barnett. *Supplemental Environmental Projects: A New Approach for EPA Enforcement*, 26 ELR 10174 (Apr. 1996).

Monsma, David & Janice Mazurek. *Building a Performance-Based Environmental Management System: EPA's Reinvention Efforts Revisited*, 6 CORP. ENVTL. STRATEGY 233 (1999).

Socolow, Robert et al. eds. INDUSTRIAL ECOLOGY AND GLOBAL CHANGE (Cambridge Univ. Press 1994).

U.S. Environmental Protection Agency. Incentives for Self-Policing: Discovery, Disclosure, Correction and Prevention of Violations, 65 Fed. Reg. 19618 (Apr. 11, 2000).

_____. Incentives for Self-Policing: Discovery, Disclosure, Correction and Prevention of Violations, 60 Fed. Reg. 66705 (Dec. 22, 1995).

Welford, Richard J. & Andrew Gouldson. ENVIRONMENTAL MANAGEMENT AND BUSINESS STRATEGY (Pitman 1993).

Environmental Auditing

Bergeson, Lynn L. & Lisa M. Campbell. *Carrot or Stick: The Debate Over Environmental Auditing*, WASH. LAW., Sept./Oct. 1997, at 38.

Casto, Keith & Tiffany Billingsley Potter. *Environmental Audits: Barriers, Opportunities and a Recommendation*, 5 HASTINGS W.-N.W. J. ENVTL. L. & POL'Y 233 (1999).

Grayson, E. Lynn & Christina M. Riewer. *EPA's Audit Policy and State Audit-Privilege Laws: Moving Beyond Command and Control?*, 27 ELR 10243 (May 1997).

Hancock, William A. CORPORATE COUNSEL'S GUIDE TO ENVIRONMENTAL COMPLIANCE AND AUDITS (Business Laws 1992).

Ledgerwood, Grant et al. IMPLEMENTING AN ENVIRONMENTAL AUDIT: HOW TO GAIN COMPETITIVE ADVANTAGE USING QUALITY AND ENVIRONMENTAL RESPONSIBILITY (Irwin 1994).

Lee, John A. & Bertram C. Frey. *Environmental Audit Immunity Laws: A State-by-State Comparison*, Env't Rep. (BNA), June 13, 1997, at 331.

Nelson, David D., ed. INTERNATIONAL ENVIRONMENTAL AUDITING (Government Institutes 1998).

Piasecki, Bruce W. et al. ENVIRONMENTAL MANAGEMENT AND BUSINESS STRATEGY: LEADERSHIP SKILLS FOR THE 21ST CENTURY (John Wiley & Sons 1999).

Shields, Jacqueline, ed. AIR EMISSIONS, BASELINES, AND ENVIRONMENTAL AUDITING (John Wiley & Sons 1997).

Sobnosky, Kevin J. *The Value-Added Benefits of Environmental Auditing*, ENVTL. QUALITY MGMT., Winter 1999, at 25.

Stoner, Nancy K. & Amanda A. Gibson. *EPA's Audit Policy Spells Success for Corporate Users, EPA, The Public, and Most Importantly, The Environment*, 29 ELR 10705 (Nov. 1999).

Stoner, Nancy K. & Wendy J. Miller. *National Conference of State Legislatures Study Finds That State Environmental Audit Laws Have No Impact on Company Self-Auditing and Disclosure of Violations*, 29 ELR 10265 (May 1999).

U.S. Department of Justice. Factors in Decisions on Criminal Prosecutions for Environmental Violations in the Context of Significant Voluntary Compliance or Disclosure Efforts by the Violator (July 1, 1999).

U.S. Environmental Protection Agency. Environmental Auditing Policy Statement, 51 Fed. Reg. 25003 (July 9, 1986).

Weaver, Grace H. STRATEGIC ENVIRONMENTAL MANAGEMENT: USING TQEM AND ISO 14000 FOR COMPETITIVE ADVANTAGE (John Wiley & Sons 1996).

Willig, John T., ed. AUDITING FOR ENVIRONMENTAL QUALITY LEADERSHIP: BEYOND COMPLIANCE TO ENVIRONMENTAL EXCELLENCE (Executive Enterprises 1995).

Environmental Management Systems

Bell, Christopher L. *How Does an ISO 14001 EMS Manage Compliance?*, INT'L ENVTL. SYS. UPDATE, Mar. 1999, at 15.

_____. *The ISO 14001 Environmental Management Systems Standard: A Modest Perspective*, 27 ELR 10622 (Dec. 1997).

_____. *Why the Government Should Stay Out of ISO*, ENVTL. F., Nov./Dec. 1997, at 24.

_____. *ISO 14001: Application of International Environmental Management Systems Standards in the United States*, 25 ELR 10678 (Dec. 1995).

Carr, Donald A. & William L. Thomas. *ISO 14001 Sets a Global Environmental Standard*, NAT'L L.J., July 8, 1996, at C2.

_____. *Devising a Compliance Strategy Under the ISO 14000 International Environmental Management Standards*, 15 PACE ENVTL. L. REV. 85 (1997).

Cascio, Joseph, ed. THE ISO 14000 HANDBOOK (American Soc'y for Quality 1999).

Coyne, Karen. *Environmental Management Systems: Plotting a Profitable Course*, ENVTL. SOLUTIONS, July 1, 1997, at 29.

Epstein, Marc J. & Marie-Joseé Roy. *Using ISO 14000 for Improved Organizational Learning and Environmental Management*, ENVTL. QUALITY MGMT., Fall 1997, at 21.

Florida, Richard. *Research Suggests EMSs, P2 Programs Favor Improved Performance*, INT'L ENVTL. SYS. UPDATE, Dec. 1999, at 14.

Hillary, Ruth, ed. ENVIRONMENTAL MANAGEMENT SYSTEMS AND CLEANER PRODUCTION (John Wiley & Sons 1997).

_____. ISO 14001: CASE STUDIES AND PRACTICAL EXPERIENCES (Greenleaf 2000).

Hunt, David & Catherine Johnson. ENVIRONMENTAL MANAGEMENT SYSTEMS: PRINCIPLES AND PRACTICE (McGraw-Hill 1995).

Morrison, Jason et al. MANAGING A BETTER ENVIRONMENT: OPPORTUNITIES AND OBSTACLES FOR ISO 14001 IN PUBLIC POLICY AND COMMERCE (Pacific Inst. 2000).

Murray, Paula C. *Inching Toward Environmental Regulatory Reform—ISO 14000: Much Ado About Nothing or a Reinvention Tool?*, 37 AM. BUS. L.J. 35 (1999).

Pezzoli, Keith. *Environmental Management Systems (EMSs) and Regulatory Innovation,* CAL. W. L. REV. (forthcoming 2000).

Reilly, Robert Anthony. *The New Paradigm: ISO 14000 and Its Place in Regulatory Reform*, 22 J. CORP. L. 535 (1997).

Room, Nigel J., ed. SUSTAINABILITY STRATEGIES FOR INDUSTRY: THE FUTURE OF CORPORATE PRACTICE (Island Press 1998).

Rosenbaum, S. Wayne. ISO AND THE LAW (AQA Co. 1998).

Rowledge, Lorinda et al. MAPPING THE JOURNEY: CASE STUDIES IN STRATEGY AND ACTION TOWARD SUSTAINABLE DEVELOPMENT (Greenleaf 1999).

Schoffman, Alan & Alan Tordini. ISO 14000: A PRACTICAL APPROACH (Oxford Univ. Press 2000).

Sheldon, Christopher, ed. ISO 14001 AND BEYOND: ENVIRONMENTAL MANAGEMENT SYSTEMS IN THE REAL WORLD (Greenleaf 1997).

Sisk, Steven W. COMPLIANCE-FOCUSED ENVIRONMENTAL MANAGEMENT SYSTEM—ENFORCEMENT AGREEMENT GUIDANCE (U.S. Environmental Protection Agency, Office of Criminal Enforcement, Forensics and Training, EPA-330/9-97-002, Aug. 1997, and as revised Jan. 1999).

Spedding, Linda S. ENVIRONMENTAL MANAGEMENT FOR BUSINESS (John Wiley & Sons 1996).

Stenzel, Paulette L. *Can the ISO 14000 Series Environmental Management Standards Provide a Viable Alternative to Government Regulation?*, 37 AM. BUS. L.J. 237 (2000).

Teter, Joni. *Environmental Management Systems in Unlikely Places: Notes From Malaysia and Yellowstone*, 29 COLORADO LAW. 67 (2000).

Tibor, Tom & Ira Feldman, eds. IMPLEMENTING ISO 14000 (Irwin 1997).

U.S. Environmental Protection Agency. Aiming for Excellence: Actions to Encourage Stewardship and Accelerate Environmental Progress (EPA 100-R-99-006, July 1999).

_____. EPA Position Statement on Environmental Management Systems and ISO 14001 and a Request for Comments on the Nature of Data to Be Collected From Environmental Management Systems/ISO 14001 Pilots, 63 Fed. Reg. 12094 (Mar. 12, 1998).

Voorhees, John. *Global Environmental Solutions: Management Systems and Synchronicity*, 28 STETSON L. REV. 1155 (1999).

Voorhees, John & Robert A. Woellner. INTERNATIONAL ENVIRONMENTAL RISK MANAGEMENT: ISO 14000 AND THE SYSTEMS APPROACH (Lewis Publishers 1998).

Welford, Richard J. CORPORATE ENVIRONMENTAL MANAGEMENT SYSTEMS AND STRATEGIES (Earthscan 1997).

Wilson, Cheryle L. & William L. Thomas. *The ASARCO-EPA Settlement Agreement: Understanding the Value of a Corporate-Wide EMS for Regulators and Strategists Alike*, CORP. ENVTL. STRATEGY, Summer 1998, at 4.

Woodside, Gayle et al. ISO 14001 IMPLEMENTATION MANUAL (McGraw-Hill 1998).

Pollution Prevention

Aldrich, James R. POLLUTION PREVENTION ECONOMICS: FINANCIAL IMPACTS ON BUSINESS AND INDUSTRY (McGraw-Hill 1995).

Becker, Monica M. & Nicholas Ashford. RECENT EXPERIENCE IN ENCOURAGING USE OF POLLUTION PREVENTION IN ENFORCEMENT SETTLEMENTS: FINAL REPORT (U.S. Environmental Protection Agency 1995).

Bergeson, Lynn L. *EPA Rolls Out a Methodology for Identifying P2 Opportunities for SEPs*, POLLUTION PREVENTION REV., Autumn 1997, at 87.

Blomquist, Robert F. *Government's Role Regarding Industrial Pollution Prevention in the United States*, 29 GA. L. REV. 349 (1995).

Cichowitz, Judith A. HOW TO CONTROL COSTS IN YOUR POLLUTION PREVENTION PROGRAM (John Wiley & Sons 1997).

Grayson, E. Lynn. *The Pollution Prevention Act of 1990: Emergence of a New Environmental Policy*, 22 ELR 10392 (June 1992).

Hopp, Rachel M. *Settling With P2—How Pollution Prevention Initiatives Can Help Your Company Correct Violations and Reduce Penalties*, POLLUTION PREVENTION REV., Autumn 1994, at 387.

Lewis, Cynthia A. & James M. Thunder. FEDERAL CHEMICAL REGULATION: TSCA, EPCRA AND THE POLLUTION PREVENTION ACT (BNA 1997).

Ochsner, Michele. *Pollution Prevention: An Overview of Regulatory Incentives and Barriers*, 6 N.Y.U. ENVTL. L.J. 586 (1998).

Rosenberg, Peter et al. *EPA's Revised SEP Policy and the Negotiation of P2 SEPs*, POLLUTION PREVENTION REV., Autumn 1995, at 1.

Strasser, Kurt A. *Cleaner Technology, Pollution Prevention and Environmental Regulation*, 9 FORDHAM ENVTL. L.J. 1 (1997).

_____. *Preventing Pollution*, 8 FORDHAM ENVTL. L.J. 1 (1996).

U.S. Environmental Protection Agency. Enforcement and Compliance Assurance, Identification of Pollution Prevention (P2) Technologies for Possible Inclusion in Enforcement Agreements Using Supplemental Environmental Projects (SEPs) and Injunctive Relief—Final Report (EPA-300-R-97-001, Mar. 1997).

Wilson, Cheryle L. *Preventing and Mitigating Environmental Enforcement With an EMS*, POLLUTION PREVENTION REV., Summer 1998, at 29.